Student Solutions Manual and Study Guide

FOR

SERWAY, VUILLE, AND FAUGHN'S

COLLEGE PHYSICS

EIGHTH EDITION, VOLUME 1

John R. Gordon
Emeritus, James Madison University

Charles Teague
Emeritus, Eastern Kentucky University

Raymond A. Serway
Emeritus, James Madison University

BROOKS/COLE
CENGAGE Learning™

Australia • Brazil • Japan • Korea • Mexico • Singapore • Spain • United Kingdom • United States

Cover Credit: © Matt Hoover, www.matthoover.com

ISBN-13: 978-0-495-55611-4

ISBN-10: 0-495-55611-4

Brooks/Cole
10 Davis Drive
Belmont, CA 94002-3098
USA

Cengage Learning is a leading provider of customized learning solutions with office locations around the globe, including Singapore, the United Kingdom, Australia, Mexico, Brazil, and Japan. Locate your local office at: **international.cengage.com/region**

Cengage Learning products are represented in Canada by Nelson Education, Ltd.

For your course and learning solutions, visit **academic.cengage.com**

Purchase any of our products at your local college store or at our preferred online store **www.ichapters.com**

Printed in the United States of America
1 2 3 4 5 6 7 11 10 09 08 07

PREFACE

This *Student Solutions Manual and Study Guide* has been written to accompany the textbook, *College Physics, Eighth Edition,* by Raymond A. Serway, Chris Vuille and Jerry S. Faughn. The purpose of this ancillary is to provide students with a convenient review of the basic concepts and applications presented in the textbook, together with solutions to selected end-of-chapter problems. This is not an attempt to rewrite the textbook in a condensed fashion. Rather, emphasis is placed upon clarifying typical troublesome points, and providing further practice in methods of problem solving.

Every textbook chapter has a matching chapter in this book and each chapter is divided into several parts. Very often, reference is made to specific equations or figures in the textbook. Each feature of this Study Guide has been included to insure that it serves as a useful supplement to the textbook. Most chapters contain the following components:

- **Notes From Selected Chapter Sections:** This is a summary of important concepts, newly defined physical quantities, and rules governing their behavior.

- **Equations and Concepts:** This is a review of the chapter, with emphasis on highlighting important concepts and describing important equations and formalisms.

- **Suggestions, Skills, and Strategies:** This section offers hints and strategies for solving typical problems that the student will often encounter in the course. In some sections, suggestions are made concerning mathematical skills that are necessary in the analysis of problems.

- **Review Checklist:** This is a list of topics and techniques the student should master after reading the chapter and working the assigned problems.

- **Solutions to Selected End-of-Chapter Problems:** Solutions are given for selected odd-numbered problems that were chosen to illustrate important concepts in each textbook chapter.

- **Tables:** A list of selected Physical Constants is printed on the inside front cover; and a table of some Conversion Factors is provided on the inside back cover.

An important note concerning significant figures: The answers to most of the end-of-chapter problems are stated to three significant figures, though calculations are carried out with as many digits as possible. We sincerely hope that this *Student Solutions Manual* *and Study Guide* will be useful to you in reviewing the material presented in the text, and in improving your ability to solve problems and score well on exams. We welcome any comments or suggestions which could help improve the content of this study guide in future editions; and we wish you success in your study.

John R. Gordon
Harrisonburg, VA

Charles Teague
Richmond, KY

Raymond A. Serway
Leesburg, VA

Acknowledgments

We are indebted to everyone who contributed to this *Student Solutions Manual and Study Guide to Accompany College Physics, Eighth Edition.*

We wish to express our sincere appreciation to the staff of Cengage Learning who provided necessary resources and coordinated all phases of this project. Special thanks go to Sylvia Krick (Assistant Freelance Editor), Stefanie Chase (Editorial Assistant), Michelle Cole (Associate Content Project Manager), Ed Dodd (Development Editor), and Chris Hall (Physics Acquisitions Editor).

Our appreciation goes to our reviewer, Ed Oberhofer, Emeritus Faculty, University of North Carolina at Charlotte. His careful reading of the manuscript and checking the accuracy of the problem solutions contributed in an important way to the quality of the final product. Any errors remaining in the manual are the responsibility of the authors.

It is a pleasure to acknowledge the excellent work of the staff of ICC Macmillan Publishing Solutions for assembling and typing this manual and preparing diagrams and page layouts. Their technical skills and attention to detail added much to the appearance and usefulness of this volume.

Finally, we express our appreciation to our families for their inspiration, patience, and encouragement.

Suggestions for Study

We have seen a lot of successful physics students. The question, "How should I study this subject?" has no single answer, but we offer some suggestions that may be useful to you.

1. Work to understand the basic concepts and principles before attempting to solve assigned problems. Carefully read the textbook before attending your lecture on that material. Jot down points that are not clear to you, take careful notes in class, and ask questions. Reduce memorization to a minimum. Memorizing sections of a text or derivations does not necessarily mean you understand the material.

2. After reading a chapter, you should be able to define any new quantities that were introduced and discuss the first principles that were used to derive fundamental equations. A review is provided in each chapter of the Study Guide for this purpose, and the marginal notes in the textbook (or the index) will help you locate these topics. You should be able to correctly associate with each physical quantity the symbol used to represent that quantity (including vector notation, if appropriate) and the SI unit in which the quantity is specified. Furthermore, you should be able to express each important principle or equation in a concise and accurate prose statement. Perhaps the best test of your understanding of the material will be your ability to answer questions and solve problems in the text, or those given on exams.

3. Try to solve plenty of the problems at the end of the chapter. The worked examples in the text will serve as a basis for your study. This Study Guide contains detailed solutions to about twelve of the problems at the end of each chapter. You will be able to check the accuracy of your calculations for any odd-numbered problem, since the answers to these are given at the back of the text.

4. Besides what you might expect to learn about physics concepts, a very valuable skill you can take away from your physics course is the ability to solve complicated problems. The way physicists approach complex situations and break them down into manageable pieces is widely useful. Starting in Section 1.10, the textbook develops a general problem-solving strategy that guides you through the steps. To help you remember the steps of the strategy, they are called *Conceptualize, Categorize, Analyze,* and *Finalize*.

General Problem-Solving Strategy

Conceptualize

- The first thing to do when approaching a problem is to *think about* and *understand* the situation. Read the problem several times until you are confident you understand what is being asked. Study carefully any diagrams, graphs, tables, or photographs that accompany the problem. Imagine a movie, running in your mind, of what happens in the problem.

- If a diagram is not provided, you should almost always make a quick drawing of the situation. Indicate any known values, perhaps in a table or directly on your sketch.

- Now focus on what algebraic or numerical information is given in the problem. In the problem statement, look for key phrases such as "starts from at rest" ($v_i = 0$), "stops" ($v_f = 0$), or "freely falls" ($a_y = -g = -9.80$ m/s^2). Key words can help simplify the problem.

- Next focus on the expected result of solving the problem. Exactly what is the question asking? Will the final result be numerical or algebraic? If it is numerical, what units will it have? If it is algebraic, what symbols will appear in it?

- Incorporate information from your own experiences and common sense. What should a reasonable answer look like? What should its order of magnitude be? You wouldn't expect to calculate the speed of an automobile to be 5×10^6 m/s.

Categorize

- Once you have a really good idea of what the problem is about, you need to *simplify* the problem. Remove the details that are not important to the solution. For example, you can often model a moving object as a particle. Key words should tell you whether you can ignore air resistance or friction between a sliding object and a surface.

- Once the problem is simplified, it is important to *categorize* the problem. How does it fit into a framework of ideas that you construct to understand the world? Is it a simple *plug-in problem,* such that numbers can be simply substituted into a definition? If so, the problem is likely to be finished when this substitution is done. If not, you face what we can call an *analysis problem*—the situation must be analyzed more deeply to reach a solution.

- If it is an analysis problem, it needs to be categorized further. Have you seen this type of problem before? Does it fall into the growing list of types of problems that you have solved previously? Being able to classify a problem can make it much easier to lay out a plan to solve it. For example, if your simplification shows that the problem can be treated as a particle moving under constant acceleration and you have already solved such a problem (such as the examples in Section 2.6), the solution to the new problem follows a similar pattern.

Analyze

- Now, you need to analyze the problem and strive for a mathematical solution. Because you have already categorized the problem, it should not be too difficult to select relevant equations that apply to the type of situation in the problem. For example, if your categorization shows that the problem involves a particle moving under constant acceleration, Equations 2.9 to 2.13 are relevant.

- Use algebra (and calculus, if necessary) to solve symbolically for the unknown variable in terms of what is given. Substitute in the appropriate numbers, calculate the result, and round it to the proper number of significant figures.

Finalize

- This final step is the most important part. Examine your numerical answer. Does it have the correct units? Does it meet your expectations from your conceptualization of the problem? What about the algebraic form of the result—before you substituted numerical values? Does it make sense? Try looking at the variables in it to see whether the answer would change in a physically meaningful way if they were drastically increased or decreased or even became zero. Looking at limiting cases to see whether they yield expected values is a very useful way to make sure that you are obtaining reasonable results.

- Think about how this problem compares with others you have done. How was it similar? In what critical ways did it differ? Why was this problem assigned? You should have learned something by doing it. Can you figure out what? Can you use your solution to expand, strengthen, or otherwise improve your framework of ideas? If it is a new category of problem, be sure you understand it so that you can use it as a model for solving future problems in the same category.

When solving complex problems, you may need to identify a series of sub-problems and apply the problem-solving strategy to each. For very simple problems, you probably don't need this whole strategy. But when you are looking at a problem and you don't know what to do next, remember the steps in the strategy and use them as a guide.

Work on problems in this Study Guide yourself and compare your solutions with ours. Your solution does not have to look just like the one presented here. A problem can sometimes be solved in different ways, starting from different principles. If you wonder about the validity of an alternative approach, ask your instructor.

5. We suggest that you use this Study Guide to review the material covered in the text, and as a guide in preparing for exams. You can use the sections Chapter Review, Notes From Selected Chapter Sections, and Equations and Concepts to focus in on any points which require further study. The main purpose of this Study Guide is to improve upon the efficiency and effectiveness of your study hours and your overall understanding of physical concepts. However, it should not be regarded as a substitute for your textbook or for individual study and practice in problem solving.

TABLE OF CONTENTS

1

Introduction

NOTES ON SELECTED CHAPTER SECTIONS

1.1 Standards of Length, Mass, and Time

Systems of units commonly used are the **SI system,** in which the units of mass, length, and time are the kilogram (kg), meter (m), and second (s), respectively; the **cgs** or **gaussian system,** in which the units of mass, length, and time are the gram (g), centimeter (cm), and second, respectively; and the **U.S. customary system,** in which the units of mass, length, and time are the slug, foot (ft), and second, respectively.

The **meter** has been redefined several times. In October 1983, it was redefined to be the distance traveled by light in a vacuum during a time of 1/299 792 458 second.

The SI unit of mass, the **kilogram,** is defined as the mass of a specific platinum-iridium alloy cylinder kept at the International Bureau of Weights and Measures at Sèvres, France.

The **second** is now defined as 9 192 631 700 times the period of one oscillation of radiation from the Cesium-133 atom.

1.2 The Building Blocks of Matter

It is useful to view the atom as a miniature solar system with a dense, positively charged nucleus occupying the position of the Sun and negatively charged electrons orbiting like the planets. Occupying the nucleus are two basic entities, protons and neutrons. The **proton** is nature's fundamental carrier of positive charge; the **neutron** has no charge and a mass about equal to that of a proton. Even more elementary building blocks than protons and neutrons exist. Protons and neutrons are each now thought to consist of three particles called **quarks.**

1.3 Dimensional Analysis

Dimensional analysis makes use of the fact that dimensions can be treated as algebraic quantities. *Quantities can be added or subtracted only if they have the same dimensions.*

1.4   Uncertainty in Measurement and Significant Figures
A **significant figure** is a reliably known digit (other than a zero that is used to locate a decimal point).

When **multiplying several quantities,** the number of significant figures in the final result is the same as the number of significant figures in the least accurate of the quantities being multiplied, where "least accurate" means "having the lowest number of significant figures." The same rule applies to division.

When **numbers are added (or subtracted),** the number of decimal places in the result should equal the smallest number of decimal places of any term in the sum (or difference).

Most of the numerical examples and end-of-chapter problems in your textbook will yield answers having either two or three significant figures.

1.5   Conversion of Units
Sometimes it is necessary to **convert units** from one system to another. A list of conversion factors can be found on the inside front cover of the ***Student Solutions Manual and Study Guide.***

1.6   Estimates and Order-of-Magnitude Calculations
Often it is useful to **estimate an answer** to a problem in which little information is given. In such a case we refer to the **order of magnitude** of a quantity, by which we mean the power of ten that is closest to the actual value of the quantity. *Usually, when an order-of-magnitude calculation is made, the results are reliable to within a factor of 10.*

1.7   Coordinate Systems
A **coordinate system** used to specify locations in space consists of:

- A fixed reference point O, called the origin

- A set of specified axes, or directions, with an appropriate scale and label on each of the axes

- Instructions that tell us how to label a point in space relative to the origin and axes

Cartesian (rectangular) coordinates and polar coordinates are two commonly used coordinate systems.

1.8   Trigonometry
You should review the basic trigonometric functions stated by Equations (1.1) and (1.2) in the **Equations and Concepts** section of this chapter.

1.9 Problem-Solving Strategy

In developing problem-solving strategies, the following steps will be helpful.

1. Read the problem carefully at least twice. Be sure you understand the nature of the problem before proceeding further.

2. Draw a suitable diagram with appropriate labels and coordinate axes, if needed.

3. Imagine what happens in the problem.

4. Identify the basic physical principle (or principles) involved, listing the knowns and unknowns.

5. Select (or derive) equation(s) as necessary to find the unknown quantities in terms of the known quantities.

6. Solve the equations for the unknowns symbolically (algebraically without substituting numerical values).

7. Substitute the given values with the appropriate units to obtain numerical values with units for the unknowns.

8. Check your answer against the following questions: Do the units match? Is the answer reasonable? Is the plus or minus sign proper or meaningful?

EQUATIONS AND CONCEPTS

The **three basic trigonometric functions** of an acute angle in a right triangle are the sine, cosine, and tangent.

$$\sin \theta = \frac{\text{side opposite to } \theta}{\text{hypotenuse}} = \frac{y}{r}$$

$$\cos \theta = \frac{\text{side adjacent to } \theta}{\text{hypotenuse}} = \frac{x}{r} \tag{1.1}$$

$$\tan \theta = \frac{\text{side opposite to } \theta}{\text{side adjacent to } \theta} = \frac{y}{x}$$

The **Pythagorean theorem** is an important relationship among the lengths of the sides of a right triangle.

$$r^2 = x^2 + y^2 \tag{1.2}$$

SUGGESTIONS, SKILLS, AND STRATEGIES

Many mathematical symbols will be used throughout this book. Some important examples are:

$\propto$	denotes	a proportionality
$<$	means	"is less than"
$>$	means	"is greater than"
$<<$	means	"is much less than"
$>>$	means	"is much greater than"
$\cong$	indicates	approximate equality
$=$	indicates	equality
$\sim$	means	"is of the order of"
Δx ("delta x")	indicates	the change in a quantity x
$|x|$	means	the absolute value of x (always positive)
Σ (capital sigma)	represents	a sum. For example,

$$x_1 + x_2 + x_3 + x_4 + x_5 = \sum_{i=1}^{5} x_i$$

REVIEW CHECKLIST

- Discuss the units and standards of SI quantities: length, mass, and time.

- Derive SI units for physical quantities (e.g., force, velocity, volume, and acceleration) from units of the three basic quantities: length, mass, and time.

- Perform a dimensional analysis of an equation containing physical quantities whose individual units are known.

- Convert units from one system to another.

- Carry out order-of-magnitude calculations or "guesstimates."

- Describe the coordinates of a point in space using both Cartesian and polar coordinate systems.

SOLUTIONS TO SELECTED END-OF-CHAPTER PROBLEMS

4. Each of the following equations was given by a student during an examination.

$$\tfrac{1}{2}mv^2 = \tfrac{1}{2}mv_0^2 + \sqrt{mgh} \qquad v = v_0 + at^2 \qquad ma = v^2$$

Do a dimensional analysis of each equation and explain why the equation can't be correct.

Solution

An equation may be incorrect for many reasons. However, if the equation is to have any possibility of being valid, it must be dimensionally correct. That is, each term in the equation must have the same dimensions as all other terms of the equation.

In the equation $\tfrac{1}{2}mv^2 = \tfrac{1}{2}mv_0^2 + \sqrt{mgh}$, the dimensions of the first two terms are

$$[mv^2] = [mv_0^2] = [m][v^2] = M\left(\frac{L}{T}\right)^2 = \frac{ML^2}{T^2}$$

The dimensions of the third term are

$$[\sqrt{mgh}] = \sqrt{[m][g][h]} = \sqrt{M\left(\frac{L}{T^2}\right)L} = \frac{M^{\frac{1}{2}}L}{T}$$

Since the dimensions of the third term are not the same as those of the first two terms, the equation is dimensionally incorrect and cannot be valid. ◊

The dimensions of the first two terms in $v = v_0 + at^2$ are

$$[v] = [v_0] = \frac{L}{T}$$

The dimensions of the third term in this equation are

$$[at^2] = [a][t^2] = \left(\frac{L}{T^2}\right)T^2 = L$$

Thus, we see that this equation is dimensionally incorrect, and hence invalid. ◊

In $ma = v^2$, the dimensions of the first term are

$$[ma] = [m][a] = M\left(\frac{L}{T^2}\right) = \frac{ML}{T^2}$$

The dimensions of the last term are

$$[v^2] = \left(\frac{L}{T}\right)^2 = \frac{L^2}{T^2}$$

Again, the equation is seen to be dimensionally incorrect, and therefore invalid. ◊

9.   How many significant figures are there in (a) 78.9 ± 0.2, (b) 3.788×10^9, (c) 2.46×10^{-6}, (d) 0.003 2?

Solution

The significant figures in a measurement include all of the digits that are known reliably plus the first digit the experimenter had to estimate.

(a)   The notation ± 0.2 indicates an uncertainty of 2 units in the first decimal place. Thus, the number 78.9 contains 2 digits that are known reliably (the 7 and the 8) and one digit with some uncertainty (the 9). Therefore, it has three significant figures.   ◊

(b)   In scientific notation, the first part of the number (3.788 in this case) is always in the range 1 to 10 and contains all of the digits serving as significant figures. The second part of the number simply gives the power of 10 that is to multiply the first part. The number of significant figures contained in 3.788×10^9 is four.   ◊

(c)   Note that the first part of the number 2.46×10^{-6} contains three digits. Thus, as discussed in (b) above, the number of significant figures contained in 2.46×10^{-6} is three.   ◊

(d)   Expressing 0.003 2 in scientific notation gives 3.2×10^{-3} (Note that in the original notation, the zeros merely serve to locate the decimal point.) Then, as discussed in part (b), observe that 0.003 2 contains two significant figures.   ◊

---

12. The radius of a circle is measured to be (10.5 ± 0.2) m. Calculate (a) the area and (b) the circumference of the circle and give the uncertainty in each value.

Solution

(a)   The area of the circle is

$$A = \pi r^2 = \pi (10.5 \text{ m} \pm 0.2 \text{ m})^2 = \pi (10.5 \text{ m} \pm 0.2 \text{ m})(10.5 \text{ m} \pm 0.2 \text{ m})$$

$$= \pi \left[(10.5 \text{ m})(10.5 \text{ m}) + (10.5 \text{ m})(\pm 0.2 \text{ m}) \right.$$

$$\left. + (\pm 0.2 \text{ m})(10.5 \text{ m}) + (\pm 0.2 \text{ m})(\pm 0.2 \text{ m}) \right]$$

$$= \pi (10.5 \text{ m})^2 \pm 2\pi (10.5 \text{ m})(0.2 \text{ m}) + \pi (0.2 \text{ m})^2$$

Since the last term in this expression is negligible in comparison to the first two terms, the area becomes

$$A = \pi (10.5 \text{ m})^2 \pm 2\pi (10.5 \text{ m})(0.2 \text{ m}) = 346 \text{ m}^2 \pm 13 \text{ m}^2 \qquad ◊$$

(b) The circumference of the circle is

$$C = 2\pi r = 2\pi(10.5 \text{ m} \pm 0.2 \text{ m}) = 2\pi(10.5 \text{ m}) \pm 2\pi(0.2 \text{ m})$$

or

$$C = 66.0 \text{ m} \pm 1.3 \text{ m} \qquad \Diamond$$

---

15. A fathom is a unit of length, usually reserved for measuring the depth of water. A fathom is approximately 6 ft in length. Take the distance from Earth to the Moon to be 250 000 miles, and use the given approximation to find the distance in fathoms.

Solution

To convert the units used to express the distance to the Moon from miles to fathoms, we use a series of ratios in which the numerator and denominator are equal to each other. Each ratio is chosen to move us one step closer to our desired units. Using this technique, we find

$$d = (250\ 000 \text{ mi})\left(\frac{5\ 280 \text{ ft}}{1 \text{ mi}}\right)\left(\frac{1 \text{ fathom}}{6 \text{ ft}}\right)$$

$$= \frac{(250\ 000)(5\ 280)}{6} \text{ fathoms} = 2 \times 10^8 \text{ fathoms} \qquad \Diamond$$

Note that the limited accuracy of the given conversion factor from fathoms to feet makes it necessary to round the final answer to one significant figure.

---

17. A firkin is an old British unit of volume equal to 9 gallons. How many cubic meters are there in 6.00 firkins?

Solution

Note that the firkin is defined to be exactly 9 gallons, so we shall not consider the factor of 9 to limit the number of significant figures that may be retained in the final answer of this conversion calculation.

As in all conversion calculations, we start with the given data and multiply it by a series of ratios, with each ratio having a value of 1 (i.e., the numerator and denominator are equal to each other). We may multiply the original data by 1 as many times as desired, and while we may change its appearance, the final result is equal to the original value. Each ratio used is chosen to cancel some of the current units and move us one step closer to the desired units.

In the calculation given below, it is desired to convert cubic inches to cubic centimeters, and later to convert cubic centimeters to cubic meters. The table of conversion factors on the front flyleaf of your textbook does not include these conversion factors. Still, this table does tell us that $1 \text{ in} = 2.54 \text{ cm}$. Thus, $(1 \text{ in})^2 = (2.54 \text{ cm})^2$ and $(1 \text{ in})^3 = (2.54 \text{ cm})^3$. This means that we may use the ratio

$$\left(\frac{2.54 \text{ cm}}{1 \text{ in}}\right)^3 = \frac{(2.54 \text{ cm})^3}{(1 \text{ in})^3} = \frac{(2.54)^3 \text{ cm}^3}{1 \text{ in}^3}$$

in our calculation. Similarly, $1 \text{ m} = 10^2 \text{ cm}$, so

$$(1 \text{ m})^2 = 1 \text{ m}^2 = (10^2 \text{ cm})^2 = 10^4 \text{ cm}^2$$

and

$$(1 \text{ m})^3 = 1 \text{ m}^3 = (10^2 \text{ cm})^3 = 10^6 \text{ cm}^3$$

The calculation needed to convert 6.00 firkins to cubic meters is

$$6.00 \text{ firkins} = 6.00 \text{ firkins} \left(\frac{9 \text{ gal}}{1 \text{ firkin}}\right)\left(\frac{231 \text{ in}^3}{1 \text{ gal}}\right)\left(\frac{(2.54)^3 \text{ cm}^3}{1 \text{ in}^3}\right)\left(\frac{1 \text{ m}^3}{10^6 \text{ cm}^3}\right)$$

$$= 0.204 \text{ m}^3 \qquad \qquad \Diamond$$

---

25. The amount of water in reservoirs is often measured in acre-ft. One acre-ft is a volume that covers an area of one acre to a depth of one foot. An acre is 43 560 ft². Find the volume in SI units of a reservoir containing 25.0 acre-ft of water.

Solution

To convert the volume of the reservoir from the given units of acre-ft to the standard units of volume in the SI system (m^3), it will be necessary to use multiple conversion factors. Each of these conversion factors will consist of a ratio which has a value of 1. The numerator and denominator of the ratio will be chosen to cancel some of the units we have and move us one step closer to the desired final units. Often, multiple conversion factors will be employed at the same time to do the entire conversion of units in a single step. We shall do the desired calculation in several steps, and then show how it could have been done in only one step.

$$V = 25.0 \text{ acre} \cdot \text{ft} = (25.0 \text{ acre} \cdot \text{ft})\left(\frac{43\ 560 \text{ ft}^2}{1 \text{ acre}}\right) = 1.09 \times 10^6 \text{ ft}^3$$

$$= 1.09 \times 10^6 \text{ ft}^3 = (1.09 \times 10^6 \text{ ft}^3)\left(\frac{1 \text{ m}}{3.281 \text{ ft}}\right) = 3.32 \times 10^5 \text{ ft}^2 \cdot \text{m}$$

$$= 3.32 \times 10^5 \text{ ft}^2 \cdot \text{m} = (3.32 \times 10^5 \text{ ft}^2 \cdot \text{m})\left(\frac{1 \text{ m}}{3.281 \text{ ft}}\right) = 1.01 \times 10^5 \text{ ft} \cdot \text{m}^2$$

$$= 1.01 \times 10^5 \text{ ft} \cdot \text{m}^2 = (1.01 \times 10^5 \text{ ft} \cdot \text{m}^2)\left(\frac{1 \text{ m}}{3.281 \text{ ft}}\right) = 3.08 \times 10^4 \text{ m}^3 \qquad \Diamond$$

To do this conversion in one step, it is helpful to realize that the ratio 1 m/3.281 ft = 1 (since the numerator and the denominator are equal). Thus, $(1 \text{ m}/3.281 \text{ ft})^3 = 1^3 = 1$. Making use of this, we write

$$V = 25.0 \text{ acre} \cdot \text{ft} = (25.0 \text{ acre} \cdot \text{ft}) \left(\frac{43\,560 \text{ ft}^2}{1 \text{ acre}} \right) \left(\frac{1 \text{ m}}{3.281 \text{ ft}} \right)^3 = 3.08 \times 10^4 \text{ m}^3 \qquad \lozenge$$

---

33. An automobile tire is rated to last for 50 000 miles. Estimate the number of revolutions the tire will make in its lifetime.

Solution

If the tire does not slip on the ground, its center moves forward a distance equal to the circumference of the tire for every revolution made.

Tires for full size automobiles often have radii of about 14 inches, or approximately 1.2 ft.

Thus, the circumference of the tire is $C = 2\pi r \approx 8$ ft, and a reasonable guess for the number of revolutions made would be

$$n = \frac{\text{distance traveled}}{\text{circumference}} = \frac{x}{2\pi r} \approx \frac{50\,000 \text{ mi}}{8 \text{ ft/rev}} \left(\frac{5\,280 \text{ ft}}{1 \text{ mi}} \right) = 3 \times 10^7 \text{ rev}$$

or

$$n \sim 10^7 \text{ rev} \qquad \lozenge$$

---

39. Two points are given in polar coordinates by $(r, \theta) = (2.00 \text{ m}, 50.0°)$ and $(r, \theta) = (5.00 \text{ m}, -50.0°)$, respectively. What is the distance between them?

Solution

Consider the sketch at the right. In this case, the first point has polar coordinates of

$$r_1 = 2.00 \text{ m}, \quad \theta_1 = 50.0°$$

and Cartesian coordinates of

$$x_1 = r_1 \cos\theta_1 = (2.00 \text{ m})\cos 50.0° = 1.29 \text{ m}$$
$$y_1 = r_1 \sin\theta_1 = (2.00 \text{ m})\sin 50.0° = 1.53 \text{ m}$$

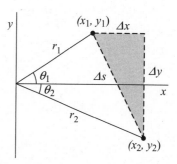

Similarly, the polar and Cartesian coordinates for the second point are:

$$r_2 = 5.00 \text{ m}, \quad \theta_2 = -50.0°$$

and

$$x_2 = r_2 \cos\theta_2 = (5.00 \text{ m})\cos(-50.0°) = 3.21 \text{ m}$$

$$y_2 = r_2 \sin\theta_2 = (5.00 \text{ m})\sin(-50.0°) = -3.83 \text{ m}$$

In moving from point 2 up to point 1, one must undergo a horizontal displacement of

$$\Delta x = x_1 - x_2 = 1.29 \text{ m} - 3.21 \text{ m} = -1.92 \text{ m}$$

and a vertical displacement of

$$\Delta y = y_1 - y_2 = 1.53 \text{ m} - (-3.83 \text{ m}) = 5.36 \text{ m}$$

Observe that these displacements are the legs of the shaded right triangle shown in the sketch. The straight line distance between the two points is the hypotenuse of this triangle and is given by the Pythagorean theorem as

$$\Delta s = \sqrt{(\Delta x)^2 + (\Delta y)^2} = \sqrt{(-1.92)^2 + (5.36)^2} = 5.69 \text{ m} \qquad ◊$$

43. A high fountain of water is located at the center of a circular pool as shown in Figure P1.43. Not wishing to get his feet wet, a student walks around the pool and measures its circumference to be 15.0 m. Next, the student stands at the edge of the pool and uses a protractor to gauge the angle of elevation at the bottom of the fountain to be 55.0°. How high is the fountain?

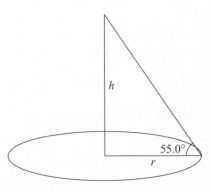

Figure P1.43

Solution

The circumference of the fountain is $C = 2\pi r$, so the radius is found to be

$$r = \frac{C}{2\pi} = \frac{15.0 \text{ m}}{2\pi} = 2.39 \text{ m}$$

Now observe the sketch given at the right and note that

$$\tan(55.0°) = \frac{h}{r} = \frac{h}{2.39 \text{ m}}$$

Solving this expression for the height of the fountain gives

$$h = (2.39 \text{ m})\tan(55.0°) = 3.41 \text{ m} \qquad ◊$$

49. A surveyor measures the distance across a straight river by the following method: Starting directly across from a tree on the opposite bank, he walks 100 m along the riverbank to establish a baseline. Then he sights across to the tree. The angle from his baseline to the tree is 35.0°. How wide is the river?

Solution

To determine the width w of the river, consider the sketch shown at the right and apply basic concepts of trigonometry.

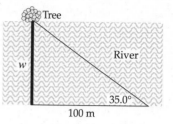

In the triangle shown, notice that the tangent of the 35.0° angle is given by

$$\tan(35.0°) = \frac{\text{opposite side}}{\text{adjacent side}} = \frac{w}{100 \text{ m}}$$

Thus, the width of the river is given by

$$w = (100 \text{ m})\tan(35.0°) = 70.0 \text{ m} \qquad \Diamond$$

55. The displacement of an object moving under uniform acceleration is some function of time and the acceleration. Suppose we write this displacement as $s = ka^m t^n$ where k is a dimensionless constant. Show by dimensional analysis that this expression is satisfied if $m = 1$ and $n = 2$. Can this analysis give the value of k?

Solution

For the equation to be valid, we must choose values of m and n to make it dimensionally consistent.

Since s is a displacement, its dimensions are those of length, $[s] = L$. The acceleration, a, is a length divided by the square of a time: $[a] = L/T^2$. The variable t has dimensions of time, $[t] = T$, and the constant k has no dimensions.

Substituting these dimensions into the proposed equation yields:

$$[s] = [a]^m [t]^n$$

$$L = \left(\frac{L}{T^2}\right)^m T^n = L^m T^{-2m} T^n$$

or

$$L^1 T^0 = L^m T^{n-2m}$$

Note that the factor T^0 introduced on the left side of the equation is equal to 1, so its introduction does not change the equation. This equation can be true only if the powers of length L are the same of the two sides of the equation and, simultaneously, the powers of time T are the same on both sides. [*Note:* If the third basic unit, mass M, were present we would also require that its powers be identical on the two sides of the equation.]

Thus, we obtain a set of two simultaneous equations:

$$1 = m \quad \text{and} \quad 0 = n - 2m$$

The solutions are seen to be:

$$m = 1 \quad \text{and} \quad n = 2 \qquad \lozenge$$

This technique gives no information about the possible values of the dimensionless constant k. $\qquad \lozenge$

61. (a) How many seconds are there in a year? (b) If one micrometeorite (a sphere with a diameter on the order of 10^{-6} m) struck each square meter of the Moon each second, estimate the number of years would it take to cover the Moon with micrometeorites to a depth of one meter. (*Hint:* Consider a cubic box, 1 m on a side, on the Moon, and find how long it would take to fill the box.)

Solution

(a) We use multiple ratios, each with the numerator equal to the denominator, to convert a time interval of 1 year to seconds.

$$1 \, \text{yr} = \left(1 \, \cancel{\text{yr}}\right)\left(\frac{365.2 \, \cancel{\text{day}}}{1 \, \cancel{\text{yr}}}\right)\left(\frac{24 \, \cancel{\text{h}}}{1 \, \cancel{\text{day}}}\right)\left(\frac{60 \, \cancel{\text{min}}}{1 \, \cancel{\text{h}}}\right)\left(\frac{60 \, \text{s}}{1 \, \cancel{\text{min}}}\right) = 3.16 \times 10^{7} \, \text{s} \qquad \lozenge$$

(b) Consider a cubical box, sunk into the Moon, having an open top with an area of $1 \, \text{m}^2$ and extending downward to a depth of 1 meter. When filled with micrometeorites, each having a diameter of 10^{-6} m, the number of meteorites along each edge of this box is

$$n = \frac{\text{length of edge}}{\text{meteorite diameter}} = \frac{1 \, \text{m}}{10^{-6} \, \text{m}} = 10^{6}$$

The total number of meteorites in the filled box is approximately

$$N \approx n \times n \times n = n^{3} = \left(10^{6}\right)^{3} = 10^{18}$$

At the rate of 1 meteorite entering the box each second, the time required to fill the box is

$$t = \frac{N}{rate} \approx \frac{10^{18}}{1 \, \text{per sec}} = \left(10^{18} \, \text{s}\right)\left(\frac{1 \, \text{yr}}{3.16 \times 10^{7} \, \text{s}}\right) = 3 \times 10^{10} \, \text{yr or } t \sim 10^{10} \, \text{yr} \qquad \lozenge$$

2

Motion in One Dimension

NOTES FROM SELECTED CHAPTER SECTIONS

2.1 Displacement

The displacement of an object, defined as its **change in position,** is given by the difference between its final and initial positions, or $x_f - x_i$. Displacement is an example of a vector quantity. *A vector is a physical quantity that requires a specification of both direction and magnitude.*

2.2 Velocity

Average speed is the ratio of the total distance traveled to the elapsed time of travel. *Speed is a scalar quantity and is always zero or positive.*

The **average velocity** of an object during the time interval t_i to t_f is equal to the slope of the straight line joining the initial and final points on a graph of the position of the object plotted versus time. *Velocity is a vector quantity and can be positive, negative, or zero.*

Instantaneous velocity (velocity at a specific time) is the slope of the line tangent to the position-time curve at a point P corresponding to the specified time.

The **instantaneous speed** of an object, which is a scalar quantity, is defined as the magnitude of the instantaneous velocity.

2.3 Acceleration

The **average acceleration** during a given time interval is defined as the change in velocity divided by the time interval during which this change occurs.

The **instantaneous acceleration** of an object at a certain time equals the slope of the tangent to the velocity vs. time graph at that instant of time.

2.4 Motion Diagrams

The **motion of an object** can be represented by a motion diagram with vectors indicating the direction and relative magnitude of the velocity and acceleration at successive time intervals. You should carefully study Active Figure 2.12 in the textbook. This figure illustrates the motion of a car in three different cases:

- Constant positive velocity, zero acceleration

- Positive velocity, positive acceleration

- Positive velocity, negative acceleration

2.5 One-Dimensional Motion with Constant Acceleration

This type of motion is important because it applies to many objects in nature. When an object moves with constant acceleration, the average acceleration equals the instantaneous acceleration. *Equations (2.6) through (2.10) may be used to solve any problem in one-dimensional motion with constant acceleration.*

2.6 Freely Falling Objects

A freely falling body is an object moving freely under the influence of gravity only, regardless of its initial motion.

It is important to emphasize that any freely falling object experiences an acceleration directed downward. *This is true regardless of the direction of motion of the object.* An object thrown upward or downward will experience the same acceleration as an object released from rest. *Once they are in free fall, all objects have an acceleration downward equal to the acceleration due to gravity.*

EQUATIONS AND CONCEPTS

The **displacement** Δx of an object is defined as its change in position. Equation (2.1) gives the displacement for a particle moving (in one-dimensional motion) from an initial position, x_i, to a final position, x_f. *For the case of one-dimensional motion, the usual vector notation is not required.*

$$\Delta x \equiv x_f - x_i \qquad (2.1)$$

The **average velocity** of an object during a time interval is the ratio of its displacement to the time interval during which the displacement occurred.

$$\bar{v} = \frac{\Delta x}{\Delta t} = \frac{x_f - x_i}{t_f - t_i} \qquad (2.2)$$

The **instantaneous velocity** is defined as the limit of the average velocity as the time interval Δt goes to zero. *Note that the instantaneous velocity of an object might have different values from instant to instant.*

$$v \equiv \lim_{\Delta t \to 0} \frac{\Delta x}{\Delta t} \qquad (2.3)$$

The **average acceleration** of an object during a time interval is the ratio of its change in velocity to the time interval during which the change in velocity occurs.

$$\bar{a} \equiv \frac{\Delta v}{\Delta t} = \frac{v_f - v_i}{t_f - t_i} \qquad (2.4)$$

The **instantaneous acceleration** is defined as the limit of the average acceleration as the time interval Δt goes to zero.

$$a \equiv \lim_{\Delta t \to 0} \frac{\Delta v}{\Delta t} \qquad (2.5)$$

The **equations of kinematics** can be used to describe one-dimensional motion with constant acceleration along the x axis. Note that each equation shows a different relationship among physical quantities: initial velocity, final velocity, acceleration, time, and displacement. Also in the form that they are shown, it is assumed that $t_i = 0$, $t_f = t$, and for convenience, $\Delta x = x - x_0$. *Remember, Equations (2.6)–(2.10) are only valid when the acceleration is constant.*

$$v = v_0 + at \qquad (2.6)$$

$$\bar{v} = \frac{v_0 + v}{2} \qquad (2.7)$$

$$\Delta x = \tfrac{1}{2}(v + v_0)t \qquad (2.8)$$

$$\Delta x = v_0 t + \tfrac{1}{2}at^2 \qquad (2.9)$$

$$v^2 = v_0^2 + 2a\Delta x \qquad (2.10)$$

The **motion of an object in free fall** (along the y-axis) can be described by Equations (2.6)–(2.10) when acceleration, a, is replaced by $-g$. *Note that since $a = -g$ in these four equations, the $+y$ axis is predefined to point upward.*

$$v = v_0 - gt$$

$$\Delta y = \tfrac{1}{2}(v + v_0)t$$

$$\Delta y = v_0 t - \tfrac{1}{2}gt^2$$

$$v^2 = v_0^2 - 2g\Delta y$$

SUGGESTIONS, SKILLS, AND STRATEGIES

The following procedure is recommended for solving problems involving motion with constant acceleration.

1. Make sure all the units in the problem are consistent. That is, if distances are measured in meters, be sure that velocities have units of m/s and accelerations have units of m/s^2.

2. Choose a coordinate system and make a labeled diagram of the problem including the directions of all displacements, velocities, and accelerations.

3. Make a list of all the quantities given in the problem and a separate list of those to be determined.

4. Select the equations (see Table 2.4), which will enable you to solve for the unknowns.

5. Construct an appropriate motion diagram and check to see if your answers are consistent with the diagram of the problem.

REVIEW CHECKLIST

- Define the displacement and average velocity of a particle in motion. Define the instantaneous velocity and understand how this quantity differs from average velocity.

- Define average acceleration and instantaneous acceleration.

- Construct a graph of displacement versus time (given a function such as $x = 5 + 3t - 2t^2$) for a particle in motion along a straight line. From this graph, you should be able to determine both average and instantaneous values of velocity by calculating the slope of the tangent to the graph.

- Apply the kinematic equations of this chapter, Equations (2.6)–(2.10), to any situation where the motion occurs under constant acceleration.

- Describe what is meant by a body in free fall (one moving under the influence of gravity—where air resistance is neglected). Recognize that the equations of constant accelerated motion apply directly to a freely falling object and that the acceleration is then given by $a = -g$ (where $g = 9.80$ m/s^2).

SOLUTIONS TO SELECTED END-OF-CHAPTER PROBLEMS

7. A motorist drives north for 35.0 minutes at 85.0 km/h and then stops for 15.0 minutes. He then continues north, traveling 130 km in 2.00 h. (a) What is his total displacement? (b) What is his average velocity?

Solution

(a) We choose northward to be the positive direction of motion. While the motorist has constant velocity $v = +85.0$ km/h, the average velocity is $v_{av} = v$ and the displacement that occurs is

$$\Delta x_1 = v(\Delta t) = \left(+85.0 \frac{km}{h}\right)\left[(35.0 \text{ min})\left(\frac{1 \text{ h}}{60 \text{ min}}\right)\right] = +49.6 \text{ km}$$

The displacement during the 15.0 min rest stop is $\Delta x_2 = 0$, and the displacement during the final 2.00 h is $\Delta x_3 = +130$ km. Thus, the total displacement for the trip is

$$\Delta x_{total} = \Delta x_1 + \Delta x_2 + \Delta x_3 = 49.6 \text{ km} + 0 + 130 \text{ km} = 180 \text{ km} \qquad \Diamond$$

(b) The duration of the trip is

$$\Delta t_{total} = \Delta t_1 + \Delta t_2 + \Delta t_3 = (35.0 \text{ min} + 15.0 \text{ min})\left(\frac{1 \text{ h}}{60 \text{ min}}\right) + 2.00 \text{ h} = 2.83 \text{ h}$$

Thus, the average velocity for the trip is

$$\bar{v} = \frac{\Delta x_{total}}{\Delta t_{total}} = \frac{+180 \text{ km}}{2.83 \text{ h}} = +63.4 \text{ km/h} \qquad \Diamond$$

13. A person takes a trip, driving with a constant speed of 89.5 km/h except for a 22.0-min rest stop. If the person's average speed is 77.8 km/h, how much time is spent on the trip and how far does the person travel?

Solution

If t_1 is the time the time the motorist travels at a constant speed of $v_1 = 89.5$ km/h, the average speed is $v_{av} = v_1$ and the displacement during this phase of the trip is

$$\Delta x_1 = v_1 t_1 = (89.5 \text{ km/h}) t_1$$

The displacement during the 22.0-min rest stop is zero, so the total displacement for the trip is

$$\Delta x_{total} = \Delta x_1 + 0 = (89.5 \text{ km/h}) t_1$$

The total time for the trip is $\Delta t_{total} = t_1 + (22.0 \text{ min}) \left(\dfrac{1 \text{ h}}{60 \text{ min}} \right) = t_1 + 0.367$ h

Thus, if the average speed for the trip is $\bar{v} = 77.8$ km/h,

$$\Delta x_{total} = \bar{v} (\Delta t_{total}) \quad \text{becomes} \quad (89.5 \text{ km/h}) t_1 = (77.8 \text{ km/h})(t_1 + 0.367 \text{ h})$$

Solving for t_1 yields $t_1 = 2.44$ h, so $\Delta t_{total} = 2.44$ h $+ 0.367$ h $= 2.80$ h. ◊

The total displacement for the trip is then

$$\Delta x_{total} = \bar{v} (\Delta t_{total}) = (77.8 \text{ km/h})(2.80 \text{ h}) = 218 \text{ km}$$ ◊

19. Runner A is initially 4.0 mi west of a flagpole and is running with a constant velocity of 6.0 mi/h due east. Runner B is initially 3.0 mi east of the flagpole and is running with a constant velocity of 5.0 mi/h due west. How far are the runners from the flagpole when they meet?

Solution

We choose a coordinate axis that has the origin at the flagpole and eastward as the positive x direction. Then, runner A is initially located at $x_{0A} = -4.0$ mi, has initial velocity $v_{0A} = +6.0$ mi/h, and zero acceleration $(a_A = 0)$. Similarly, runner B is initially located at $x_{0B} = +3.0$ mi, has initial velocity $v_{0B} = -5.0$ mi/h, and zero acceleration $(a_B = 0)$.

The kinematics equation $\Delta x = x - x_0 = v_0 t + \frac{1}{2} a t$ may be used to obtain each runner's position at time t. This gives

Runner A: $\quad x_A - (-4.0 \text{ mi}) = (6.0 \text{ mi/h}) t + 0 \quad$ or $\quad x_A = -4.0 \text{ mi} + (6.0 \text{ mi/h}) t$

Runner B: $\quad x_B - (+3.0 \text{ mi}) = (-5.0 \text{ mi/h}) t + 0 \quad$ or $\quad x_B = 3.0 \text{ mi} - (5.0 \text{ mi/h}) t$

When the two runners meet, their x coordinates will be the same $(x_A = x_B)$. Requiring that this be true and solving the resulting equation for t will give the elapsed time when they meet:

$$x_A = x_B \quad \Rightarrow \quad -4.0 \text{ mi} + (6.0 \text{ mi/h})t = 3.0 \text{ mi} - (5.0 \text{ mi/h})t$$

$$\text{or} \quad (6.0 \text{ mi/h} + 5.0 \text{ mi/h})t = 3.0 \text{ mi} + 4.0 \text{ mi}$$

$$\text{and} \quad t = \frac{7.0 \text{ mi}}{11 \text{ mi/h}} = 0.64 \text{ h}$$

At this time,

$$x_A = -4.0 \text{ mi} + (6.0 \text{ mi/h})(0.64 \text{ h}) = -0.18 \text{ mi}$$

and

$$x_B = 3.0 \text{ mi} - (5.0 \text{ mi/h})(0.64 \text{ h}) = -0.18 \text{ mi}$$

Thus, the runners meet 0.18 mi west of the flagpole. ◊

---

25. A steam catapult launches a jet aircraft from the aircraft carrier *John C. Stennis,* giving it a speed of 175 mi/h in 2.50 s. (a) Find the average acceleration of the plane. (b) Assuming the acceleration is constant, find the distance the plane moves.

Solution

(a) The plane starts from rest $(v_0 = 0)$ and reaches a velocity of $v = 175 \text{ mi/h}$ in an elapsed time of $\Delta t = 2.50 \text{ s}$.

The average acceleration over a time interval is defined as

$$\bar{a} = \frac{\Delta v}{\Delta t} = \frac{v - v_0}{\Delta t}$$

Thus, the average acceleration of this plane during launch is

$$\bar{a} = \frac{175 \text{ mi/h} - 0}{2.50 \text{ s}} = 70.0 \; \frac{\text{mi}}{\text{h} \cdot \text{s}} \qquad ◊$$

or

$$\bar{a} = 70.0 \; \frac{\text{mi}}{\text{h} \cdot \text{s}} \left(\frac{1609 \text{ m}}{1 \text{ mi}} \right) \left(\frac{1 \text{ h}}{3600 \text{ s}} \right) = 31.3 \; \text{m/s}^2 \qquad ◊$$

and

$$\bar{a} = 31.3 \; \frac{\text{m}}{\text{s}^2} \left(\frac{1 \text{ g}}{9.80 \text{ m/s}^2} \right) = 3.19 \text{ g} \qquad ◊$$

(b) If the acceleration is constant, the displacement of the plane in time Δt is given by $\Delta x = v_0(\Delta t) + \frac{1}{2}a(\Delta t)^2$. Assuming that the plane's acceleration is constant during launch, the distance it moves in the process is then

$$\Delta x = 0 + \frac{1}{2}\left(31.3 \ \frac{m}{s^2}\right)(2.50 \ s)^2 = 97.8 \ m \qquad \Diamond$$

or

$$\Delta x = 97.8 \ m\left(\frac{3.281 \ ft}{1 \ m}\right) = 321 \ ft \qquad \Diamond$$

35. Speedy Sue, driving at 30.0 m/s, enters a one-lane tunnel. She then observes a slow-moving van 155 m ahead traveling at 5.00 m/s. Sue applies her brakes but can accelerate only at -2.00 m/s^2 because the road is wet. Will there be a collision? State how you decide. If yes, determine how far into the tunnel and at what time the collision occurs. If no, determine the distance of closest approach between Sue's car and the van.

Solution

Choose $x = 0$ and $t = 0$ at the location of Sue's car when she first spots the van and applies the brakes, and take the positive direction to be the direction the car is traveling. Then, the initial conditions for Sue's car are $x_{0S} = 0$, and $v_{0S} = +30.0$ m/s. The initial conditions for the van are $x_{0V} = +155$ m, and $v_{0V} = +5.00$ m/s. For times $t > 0$, the constant accelerations of the two vehicles are $a_S = -2.00$ m/s^2 and $a_V = 0$.

These initial conditions and the kinematics equation $\Delta x = x - x_0 = v_0 t + \frac{1}{2}at^2$ will give the positions of each vehicle as functions of time.

$$\text{Sue's Car:} \qquad x_S - 0 = (30.0 \ \text{m/s})t + \frac{1}{2}(-2.00 \ \text{m/s}^2)t^2$$

$$\text{or} \qquad x_S = (30.0 \ \text{m/s})t - (1.00 \ \text{m/s}^2)t^2$$

$$\text{Van:} \qquad x_V - 155 \ \text{m} = (5.00 \ \text{m/s})t + \frac{1}{2}(0)t^2$$

$$\text{or} \qquad x_V = 155 \ \text{m} + (5.00 \ \text{m/s})t$$

If a collision is to occur, the two vehicles must be at the same location $(x_S = x_V)$ at some time. Thus, we test for a collision by equating the equations for the x coordinates and see if the resulting equations has any real solutions.

$$x_S = x_V: \quad (30.0 \ \text{m/s})t - (1.00 \ \text{m/s}^2)t^2 = 155 \ \text{m} + (5.00 \ \text{m/s})t$$

or

$$(1.00 \ \text{m/s}^2)t^2 + (-25.0 \ \text{m/s})t + 155 \ \text{m} = 0$$

The quadratic formula then gives

$$t = \frac{-(-25.0 \text{ m/s}) \pm \sqrt{(-25.0 \text{ m/s})^2 - 4(1.00 \text{ m/s}^2)(155 \text{ m})}}{2(1.00 \text{ m/s}^2)}$$

or

$$t = 13.6 \text{ s (using upper sign) and } t = 11.4 \text{ s (using lower sign)}$$

These solutions are real, not imaginary, so a collision will occur! ◊

The smaller of the above solutions is the actual collision time. The larger solution tells when the van would again come even with the car (and start to pull ahead) if the two vehicles could pass harmlessly through each other. The location of the collision (measured from the start of the tunnel where Sue first spotted the van) is

$$x_S\big|_{t=11.4 \text{ s}} = x_V\big|_{t=11.4 \text{ s}} = 155 \text{ m} + (5.00 \text{ m/s})(11.4 \text{ s}) = 212 \text{ m} \qquad ◊$$

---

39. A car starts from rest and travels for 5.0 s with a uniform acceleration of +1.5 m/s². The driver then applies the brakes, causing a uniform acceleration of –2.0 m/s². If the brakes are applied for 3.0 s, (a) how fast is the car going at the end of the braking period, and (b) how far has the car gone?

Solution

During the first acceleration period of duration $\Delta t_1 = 5.0$ s, the car starts from rest $(v_{01} = 0)$ and accelerates at $a_1 = +1.5$ m/s². The velocity of the car at the end of this period is

$$v_{f1} = v_{01} + a_1(\Delta t_1) = 0 + (1.5 \text{ m/s}^2)(5.0 \text{ s}) = 7.5 \text{ m/s}$$

This is also the velocity of the car at the start of the braking period $(v_{02} = v_{f1} = 7.5 \text{ m/s})$.

(a) During the braking period, the acceleration is $a_2 = -2.0$ m/s² and $\Delta t_2 = 3.0$ s. The velocity at the end of this period will be

$$v_{f2} = v_{02} + a_2(\Delta t_2) = 7.5 \text{ m/s} + (-2.0 \text{ m/s}^2)(3.0 \text{ s}) = 1.5 \text{ m/s} \qquad ◊$$

(b) The total distance traveled during the two periods is

$$\Delta x_{total} = \Delta x_1 + \Delta x_2 = \bar{v}_1 \Delta t_1 + \bar{v}_2 \Delta t_2 = \left(\frac{v_{f1} + v_{01}}{2}\right)\Delta t_1 + \left(\frac{v_{f2} + v_{02}}{2}\right)\Delta t_2$$

or

$$\Delta x_{total} = \left(\frac{7.5 \text{ m/s} + 0}{2}\right)(5.0 \text{ s}) + \left(\frac{1.5 \text{ m/s} + 7.5 \text{ m/s}}{2}\right)(3.0 \text{ s}) = 32 \text{ m} \qquad ◊$$

43.   A hockey player is standing on his skates on a frozen pond when an opposing player, moving with a uniform speed of 12 m/s, skates by with the puck. After 3.0 s, the first player makes up his mind to chase his opponent. If he accelerates uniformly at 4.0 m/s², (a) how long does it take him to catch his opponent, and (b) how far has he traveled in that time? (Assume the player with the puck remains in motion at constant speed.)

Solution

(a)   Choose $x = 0$ at the initial location of the player, and $t = 0$ to be the instant when the player starts to chase his opponent. At this time, the opponent is 36 m in front of the player. At time $t > 0$, the displacements of the players from the origin are

$$x_{player} = \left(x_0\right)_{player} + \left(v_0\right)_{player} t + \frac{1}{2} a_{player} t^2 = 0 + 0 + \frac{1}{2}\left(4.0 \text{ m/s}^2\right)t^2 \tag{1}$$

and

$$x_{opponent} = \left(x_0\right)_{opponent} + \left(v_0\right)_{opponent} t + \frac{1}{2} a_{opponent} t^2 = 36 \text{ m} + \left(12 \text{ m/s}\right)t + 0 \tag{2}$$

When the players are side-by-side,

$$x_{player} = x_{opponent} \tag{3}$$

Substituting Equations (1) and (2) into Equation (3) gives

$$\frac{1}{2}\left(4.0 \text{ m/s}^2\right)t^2 = \left(12 \text{ m/s}\right)t + 36 \text{ m} \qquad \text{or} \qquad t^2 - \left(6.0 \text{ s}\right)t - 18 \text{ s}^2 = 0$$

The quadratic formula then yields

$$t = \frac{-\left(-6.0 \text{ s}\right) \pm \sqrt{\left(-6.0 \text{ s}\right)^2 - 4(1)\left(-18 \text{ s}^2\right)}}{2(1)}$$

with solutions of $t = -2.2$ s and $t = +8.2$ s. Since the time must be greater than zero, we must choose the positive solution as the time required for the player to overtake his opponent.

$$t = 8.2 \text{ s} \qquad\qquad \Diamond$$

(b)   From Equation (1), the displacement the player undergoes while chasing his opponent is

$$x_{player} = \frac{1}{2} a_{player} t^2 = \frac{1}{2}\left(4.0 \text{ m/s}^2\right)\left(8.2 \text{ s}\right)^2 = 1.3 \times 10^2 \text{ m} \qquad \Diamond$$

47. A certain freely falling object requires 1.50 s to travel the last 30.0 m before it hits the ground. From what height above the ground did it fall?

Solution

During the last interval $(\Delta t_2 = 1.50 \text{ s})$ of its drop, the object has displacement $\Delta y_2 = -30.0$ m, acceleration $a_2 = -g = -9.80 \text{ m/s}^2$, and initial velocity $v_{02} = v_{f1}$ (here v_{f1} is the velocity the object gained in falling from rest to a point 30.0 m above the ground). This initial velocity, v_{02}, may be determined by

$$\Delta y_2 = v_{02}(\Delta t_2) + \tfrac{1}{2}a_2(\Delta t_2)^2$$

or

$$v_{02} = \frac{\Delta y_2 - \tfrac{1}{2}a_2(\Delta t_2)^2}{(\Delta t_2)} = \frac{\Delta y_2}{(\Delta t_2)} - \frac{1}{2}a_2(\Delta t_2)$$

This gives

$$v_{02} = \frac{-30.0 \text{ m}}{1.50 \text{ s}} - \frac{1}{2}(-9.80 \text{ m/s}^2)(1.50 \text{ s}) = -12.7 \text{ m/s}$$

The displacement the object must have undergone during the first interval of its drop in order to start from rest $(v_{01} = 0)$ and reach this velocity $(v_{f1} = v_{02} = -12.7 \text{ m/s})$, while falling freely with acceleration $a_1 = -g = -9.80 \text{ m/s}^2$, is found from $v^2 = v_0^2 + 2a(\Delta y)$ as

$$\Delta y_1 = \frac{v_{f1}^2 - v_{01}^2}{2a_1} = \frac{(-12.7 \text{ m/s})^2 - 0}{2(-9.80 \text{ m/s}^2)} = -8.23 \text{ m}$$

The total distance the object falls before it reaches the ground is then

$$d = |\Delta y_1| + |\Delta y_2| = |-8.23 \text{ m}| + |-30.0 \text{ m}| = 38.2 \text{ m} \qquad \lozenge$$

53. A model rocket is launched straight upward with an initial speed of 50.0 m/s. It accelerates with a constant upward acceleration of 2.00 m/s² until its engines stop at an altitude of 150 m. (a) What can you say about the motion of the rocket after its engines stop? (b) What is the maximum height reached by the rocket? (c) How long after liftoff does the rocket reach its maximum height? (d) How long is the rocket in the air?

Solution

(a) After the engines stop, the rocket is a freely falling body. It continues upward for a while, slowing under the influence of gravity until it comes to rest momentarily at its maximum altitude. It then falls back to Earth, gaining speed as it falls. $\qquad \lozenge$

(b) While the engines are running, the rocket undergoes a displacement of $\Delta y_1 = +150$ m, has an acceleration of $a_1 = +2.00$ m/s^2, and had an initial velocity of $v_{01} = +50.0$ m/s. The velocity of the rocket (v_{f1}) when the engines shut off may be found from $v^2 = v_0^2 + 2a(\Delta y)$ as

$$v_{f1} = +\sqrt{v_{01}^2 + 2a_1(\Delta y_1)} = +\sqrt{(50.0 \text{ m/s})^2 + 2(+2.00 \text{ m/s}^2)(150 \text{ m})} = 55.7 \text{ m/s}$$

This is also the initial velocity $(v_{02} = v_{f1})$ for the coasting phase of the rocket, which ends when the rocket comes to rest momentarily $(v_{f2} = 0)$ at the maximum altitude. During the coasting phase, the rocket has acceleration $a_2 = -g$, so $v^2 = v_0^2 + 2a(\Delta y)$ gives the displacement while coasting as

$$\Delta y_2 = \frac{v_{f2}^2 - v_{02}^2}{2(-g)} = \frac{0 - (55.7 \text{ m/s})^2}{2(-9.80 \text{ m/s}^2)} = 158 \text{ m}$$

The maximum height reached is $h_{\max} = \Delta y_1 + \Delta y_2 = 150 \text{ m} + 158 \text{ m} = 308 \text{ m}$.   ◊

(c) The time that the engines ran is given by

$$\Delta t_1 = \frac{\Delta v_1}{a_1} = \frac{v_{f1} - v_{01}}{a_1} = \frac{55.7 \text{ m/s} - 50.0 \text{ m/s}}{2.00 \text{ m/s}^2} = 2.84 \text{ s}$$

Similarly, the duration of the coasting phase of the trip is

$$\Delta t_2 = \frac{\Delta v_2}{a_2} = \frac{v_{f2} - v_{02}}{a_2} = \frac{0 - 55.7 \text{ m/s}}{-9.80 \text{ m/s}^2} = 5.67 \text{ s}$$

So the total time the rocket takes to reach the maximum height is

$$\Delta t_{up} = \Delta t_1 + \Delta t_2 = 2.84 \text{ s} + 5.67 \text{ s} = 8.51 \text{ s}$$ ◊

(d) The time required for the rocket to fall to the ground, starting from rest $(v_{03} = 0)$ at an attitude of 308 m (so $\Delta y_3 = -308$ m), with acceleration $a_3 = -g$ is given by $\Delta y_3 = v_{03}(\Delta t_{down}) + \frac{1}{2} a_3 (\Delta t_{down})^2$ as

$$-308 \text{ m} = 0 + \frac{1}{2}(-9.80 \text{ m/s}^2)(\Delta t_{down})^2$$

or

$$\Delta t_{down} = \sqrt{\frac{2(-308 \text{ m})}{-9.80 \text{ m/s}^2}} = 7.93 \text{ s}$$

The total time the rocket is in the air is

$$\Delta t_{total} = \Delta t_{up} + \Delta t_{down} = 8.51 \text{ s} + 7.93 \text{ s} = 16.4 \text{ s}$$ ◊

59. A student throws a set of keys vertically upward to his fraternity brother, who is in a window 4.00 m above. The brother's outstretched hand catches the keys 1.50 s later. (a) With what initial velocity were the keys thrown? (b) What was the velocity of the keys just before they were caught?

Solution

Taking upward as the positive vertical direction, the keys have a constant acceleration of $a = -9.80 \text{ m/s}^2$ and undergo an upward displacement of $\Delta y = +4.00$ m during the 1.50-s time interval.

(a) Thus, $\Delta y = v_0 t + \dfrac{1}{2} a t^2$ gives the velocity at the beginning of this interval as

$$v_0 = \frac{\Delta y - \frac{1}{2} a t^2}{t} = \frac{4.00 \text{ m} - \frac{1}{2}\left(-9.80 \text{ m/s}^2\right)\left(1.50 \text{ s}\right)^2}{1.50 \text{ s}} = +10.0 \text{ m/s}$$

$$v_0 = 10.0 \text{ m/s } \mathbf{upward} \qquad \Diamond$$

(b) The velocity of the keys 1.50 s later (that is, just before they are caught) is

$$v = v_0 + at = +10.0 \text{ m/s} + \left(-9.80 \text{ m/s}^2\right)\left(1.50 \text{ s}\right) = -4.68 \text{ m/s},$$

or

$$v = 4.68 \text{ m/s } \mathbf{downward} \qquad \Diamond$$

---

63. A ball is thrown upward from the ground with an initial speed of 25 m/s; at the same instant, another ball is dropped from a building 15 m high. After how long will the balls be at the same height?

Solution

Choose upward as the positive vertical direction. Then, after the balls are released, they are both freely falling objects with acceleration $a = -g$.

During the time interval from the moment the two balls are released until the instant they are at the same height h above the ground, the ball thrown upward undergoes a displacement $\Delta y_1 = y - y_0 = +h - 0$. Thus, $\Delta y = v_0 t + \frac{1}{2} at^2$ gives

$$h = \left(+25 \text{ m/s}\right)t - \frac{1}{2}gt^2 \tag{1}$$

For this same time interval, the displacement of the dropped ball is $\Delta y_2 = y - y_0 = h - 15 \text{ m}$ and $\Delta y = v_0 t + \frac{1}{2} at^2$ becomes

$$h - 15 \text{ m} = (0)t + \frac{1}{2}(-g)t^2 \quad \text{or} \quad h = 15 \text{ m} - \frac{1}{2}gt^2 \tag{2}$$

Substituting from Equation (1) for h in Equation (2) yields

$$(25 \text{ m/s})t - \frac{1}{2}gt^2 = 15 \text{ m} - \frac{1}{2}gt^2 \quad \text{or} \quad t = \frac{15 \text{ m}}{25 \text{ m/s}} = 0.60 \text{ s} \qquad \Diamond$$

---

70. A mountain climber stands at the top of a 50.0-m cliff that overhangs a calm pool of water. She throws two stones vertically downward 1.00 s apart and observes that they cause a single splash. The first stone had an initial velocity of −2.00 m/s. (a) How long after release of the first stone did the two stones hit the water? (b) What initial velocity must the second stone have had, given that they hit the water simultaneously? (c) What was the velocity of each stone at the instant it hit the water?

Solution

Both stones are freely falling objects, with acceleration $a = -g = -9.80 \text{ m/s}^2$, from the instant they leave the thrower's hand until the instant they hit the water. Also, the displacement of each stone during the downward trip is $\Delta y = -50.0 \text{ m}$.

(a) Using $v^2 = v_0^2 + 2a(\Delta y)$, the velocity of the first stone when it reaches the water is found to be

$$v_1 = -\sqrt{(v_0)_1^2 + 2a(\Delta y)} = -\sqrt{(-2.00 \text{ m/s})^2 + 2(-9.80 \text{ m/s}^2)(-50.0 \text{ m})} = -31.4 \text{ m/s}$$

The time of flight of this stone is given by $v = v_0 + at$ as

$$t_1 = \frac{v_1 - (v_0)_1}{a} = \frac{-31.4 \text{ m/s} - (-2.00 \text{ m/s})}{-9.80 \text{ m/s}^2} = 3.00 \text{ s} \qquad \Diamond$$

(b) Since they strike the water simultaneously, the time of flight of the second stone (released 1.00 s after the first stone) is

$$t_2 = t_1 - 1.00 \text{ s} = 3.00 \text{ s} - 1.00 \text{ s} = 2.00 \text{ s}$$

Thus, $\Delta y = v_0 t + \frac{1}{2}at^2$ gives the initial velocity of the second stone as

$$(v_0)_2 = \frac{\Delta y - \frac{1}{2}at_2^2}{t_2} = \frac{-50.0 \text{ m} - \frac{1}{2}(-9.80 \text{ m/s}^2)(2.00 \text{ s})^2}{2.00 \text{ s}} = -15.2 \text{ m/s} \qquad \Diamond$$

(c) From part (a), the final velocity of the first stone is $v_1 = -31.4$ m/s. $\qquad \Diamond$

Also, $v_2 = (v_0)_2 + at_2 = -15.2 \text{ m/s} + (-9.80 \text{ m/s}^2)(2.00 \text{ s}) = -34.8 \text{ m/s}$ $\qquad \Diamond$

3

Vectors and Two-Dimensional Motion

NOTES FROM SELECTED CHAPTER SECTIONS

3.1 Vectors and Their Properties

3.2 Components of a Vector

A **scalar** has only magnitude and no direction. On the other hand, a **vector** is a physical quantity that requires the specification of both direction and magnitude. In Cartesian coordinates the **components of a vector** are the projections of the vector along the axes of a rectangular coordinate system. *A vector can be completely described by its components.*

Equal vectors have both the same magnitude and direction. When two or more vectors are **added,** they must have the same units. Two vectors that are the **negative** of each other have the same magnitude but opposite directions. **Multiplication** by a positive (negative) scalar, results is a vector in the same (opposite) direction, with a magnitude equal to the product of the scalar and the magnitude of the original vector.

3.3 Displacement, Velocity, and Acceleration in Two Dimensions

Consider an object moving between two points in space. The **displacement** of the object is defined as the change in its position vector, $\Delta \vec{r}$. The **average velocity** of a particle during the time interval Δt is the ratio of its displacement to the time interval for this displacement. The average velocity is a vector quantity directed along $\Delta \vec{r}$. The **instantaneous velocity $\vec{v}$** is defined as the limit of the average velocity as Δt goes to zero. The direction of the instantaneous velocity vector is along a line that is tangent to the path of the particle and in the direction of motion. The **average acceleration** of an object whose velocity changes is the ratio of the net change in velocity to the time interval during which the change occurs.

In a given time interval, a particle can accelerate in several ways: the magnitude of the velocity vector (the speed) may change; the direction of the velocity vector may change, making a curved path, even though the speed is constant; or both the magnitude and direction of the velocity vector may change.

3.4 Motion in Two Dimensions

In the case of **projectile motion,** if it is assumed that air resistance is negligible and that the rotation of the Earth does not affect the motion, then the motion has the following characteristics:

- The horizontal component of velocity, v_x, remains constant because there is no horizontal component of acceleration.

- The vertical component of acceleration is equal to the acceleration due to gravity.

- The vertical component of velocity, v_y, and the displacement in the y-direction are identical to those of a freely falling body.

- Projectile motion can be described as a superposition of the two motions in the x- and y-directions. *The horizontal and vertical components of the motion of a projectile are completely independent of each other.*

Review the procedure recommended in the Suggestions, Skills, and Strategies section for solving projectile motion problems.

3.5 Relative Velocity

Observations made by observers in **different frames of reference** can be related to each other using the techniques of transformation of relative velocities.

When two objects are each moving with respect to a stationary reference frame (e.g., the Earth), each moving object has a **relative velocity** with respect to the other one. You must use vector addition in order to determine the relative velocity of one object with respect to another. See the suggestions for solving relative velocity problems in the Suggestions, Skills, and Strategies section.

EQUATIONS AND CONCEPTS

The **commutative law of addition** states that when two or more vectors are added, the resultant is independent of the order of addition.

$$\vec{A} + \vec{B} = \vec{B} + \vec{A}$$

The **associative law of addition** states that when three or more vectors are added, the sum is independent of the manner in which the individual vectors are grouped.

$$\vec{A} + (\vec{B} + \vec{C}) = (\vec{A} + \vec{B}) + \vec{C}$$

In the **graphical or geometric method of vector addition,** the vectors to be added are represented by arrows connected head-to-tail *in any order* and the **resultant or sum** is the vector that joins the tail of the first vector to the head of the last vector. *The length of each vector must correspond to the magnitude of the vector according to a chosen scale. The direction of each vector must be along a direction which makes the proper angle relative to the others. All vectors in a sum must represent the same physical quantity (i.e., have the same units).*

$$\vec{R} = \vec{A} + \vec{B} + \vec{C} + \vec{D}$$

The **operation of vector subtraction** utilizes the definition of the negative of a vector. The negative of vector $\vec{A}$ is the vector that has a magnitude equal to the magnitude of $\vec{A}$ but acts or points along a direction opposite the direction of $\vec{A}$.

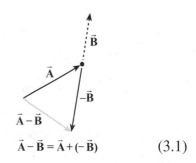

$$\vec{A} - \vec{B} = \vec{A} + (-\vec{B}) \tag{3.1}$$

The **rectangular components of a vector** are the projections of the vector onto the respective coordinate axes. The projection of $\vec{A}$ onto the x-axis, A_x, is the x-component of $\vec{A}$; and the projection of $\vec{A}$ onto the y-axis, A_y, is the y-component of $\vec{A}$. *The vector components will be positive or negative depending on the value of the angle θ which is measured counterclockwise relative to the positive x-axis.*

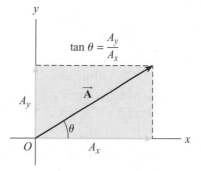

$$A_x = A\cos\theta \tag{3.2}$$
$$A_y = A\sin\theta$$

The **magnitude of vector $\vec{A}$ and the angle** which the vector makes with the positive x-axis can be determined from the values of the x- and y-components of $\vec{A}$.

$$A = \sqrt{A_x^2 + A_y^2} \tag{3.3}$$

$$\tan\theta = \frac{A_y}{A_x} \tag{3.4}$$

The **path of a projectile** is curved in the shape of a parabola as shown in the figure below. The vector that represents the initial velocity, $\vec{v}_0$, makes an angle of θ_0 (called the projection angle or angle of launch) with the horizontal. In order to analyze projectile motion, you should separate the motion into two parts: the x-components (horizontal motion) and the y-components (vertical motion). Then apply the equations of constant acceleration to each part separately.

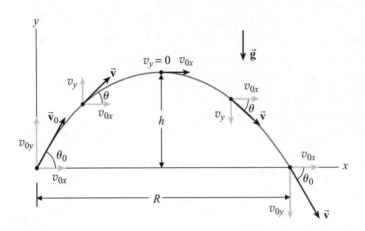

Velocity components, maximum height, and range in projectile motion

Maximum height of a projectile (h) can be found by combining Equations (3.14a) and (3.14b).

$$h = \frac{v_0^2 \sin^2 \theta_0}{2g}$$

Range of a projectile (R) can be found by combining Equations (3.13b) and (3.14b). In each case, t is eliminated between the respective pair of equations. Note that $\Delta y = h$ when $v_y = 0$ and $\Delta x = R$ when $\Delta y = 0$.

$$R = \frac{v_0^2}{g} \sin(2\theta_0)$$

The **initial horizontal and vertical components of velocity** of a projectile depend on the magnitude of the initial velocity vector and the initial angle of launch.

$$v_{0x} = v_0 \cos\theta_0$$
$$v_{0y} = v_0 \sin\theta_0$$

The horizontal component of velocity for a projectile remains constant $\left(a_x = 0\right)$; **the vertical component** decreases uniformly with time $\left(a_y = -g\right)$.

$$v_x = v_{0x} = v_0 \cos\theta_0 = \text{constant} \qquad (3.13a)$$
$$v_y = v_0 \sin\theta_0 - gt \qquad (3.14a)$$

The x- and y-coordinates of the position of a projectile are functions of the elapsed time. *In Equations* (3.14a) *and* (3.14b), *the positive direction for the vertical motion is assumed to be upward.*

$$\Delta x = v_{0x}t = \left(v_0 \cos\theta_0\right)t \qquad (3.13b)$$
$$\Delta y = (v_0 \sin\theta_0)t - \tfrac{1}{2}gt^2 \qquad (3.14b)$$

The **velocity in the y direction** can be related to the initial velocity and the displacement as stated in Equation (3.14c).

$$v_y^2 = (v_o \sin\theta_o)^2 - 2g\Delta y \qquad (3.14c)$$

The **magnitude and direction of the velocity vector** at any time can be determined from the values of v_x and v_y.

$$v = \sqrt{v_x^2 + v_y^2}$$

$$\theta = \tan^{-1}\left(\frac{v_y}{v_x}\right)$$

Relative velocity must be considered when both an observer and an object whose velocity is to be measured are moving with respect to a fixed frame of reference.

$$\vec{\mathbf{v}}_{AB} = \vec{\mathbf{v}}_{AE} - \vec{\mathbf{v}}_{BE} \qquad (3.16)$$

$\vec{\mathbf{v}}_{AB}$ = velocity of A relative to B

$\vec{\mathbf{v}}_{AE}$ = velocity A relative to the fixed frame

$\vec{\mathbf{v}}_{BE}$ = velocity of B relative to the fixed frame

SUGGESTIONS, SKILLS, AND STRATEGIES

ADDITION AND SUBTRACTION OF VECTORS

When two or more vectors are to be added, the following step-by-step procedure is recommended:

1. Select a coordinate system.

2. Draw a sketch of the vectors to be added (or subtracted), with a label on each vector.

3. Find the x- and y-components of all vectors.

4. Find the algebraic sum of the x-components of the vectors, and the algebraic sum of the y-components of the vectors. These two sums are the x- and y-components of the resultant vector.

5. Use the Pythagorean theorem to find the magnitude of the resultant vector.

6. Use a suitable trigonometric function—e.g., Equation (3.4)—to find the angle the resultant vector makes with the x-axis.

SOLVING PROJECTILE MOTION PROBLEMS

The following procedure is recommended for solving projectile motion problems:

1. Sketch the path of the projectile on a set of coordinate axes. Include the initial velocity vector and the projectile angle.

2. Resolve the initial velocity vector into x- and y-components.

3. Treat the horizontal motion (zero acceleration) and the vertical motion (acceleration $= -g$) independently.

4. Follow the techniques for solving problems with constant velocity (zero acceleration) to analyze the horizontal motion of the projectile.

5. Follow the techniques for solving problems with constant acceleration to analyze the vertical motion of the projectile.

SOLVING RELATIVE VELOCITY PROBLEMS

1. Examine the statement of the problem for phrases like ". . . the velocity of A relative to B. . . ." When a velocity is not explicitly stated as being relative to a specific observer, it is usually relative to the Earth.

2. Show the velocity vectors in a diagram; label each vector involved (usually two) with a letter that reminds you what it represents (i.e., E for Earth).

3. Assemble the three velocities you have identified into a vector equation with subscripts that represent the relationship among the velocities. For example, "the velocity of observer A relative to observer B equals the velocity of observer A relative to the Earth minus the velocity of observer B relative to the Earth" would appear as $\vec{v}_{AB} = \vec{v}_{AE} - \vec{v}_{BE}$.

4. Solve the equation in step 3 for the unknown velocity. In the case of one-dimensional motion, this will involve one equation and one unknown; for two-dimensional motion there will be two component equations and two unknowns.

REVIEW CHECKLIST

- Understand and describe the basic properties of vectors, resolve a vector into its rectangular components, use the rules of vector addition (including graphical solutions for addition of two or more vectors), and determine the magnitude and direction of a vector from its rectangular components.

- Sketch a typical trajectory of a particle moving in the xy-plane and draw vectors to illustrate the manner in which the displacement, velocity, and acceleration of the particle change with time.

- Recognize that two-dimensional motion in the xy-plane with constant acceleration is equivalent to two independent motions: constant velocity along the x-direction and constant acceleration along the y-direction.

- Recognize the fact that if the initial speed and initial angle of a projectile motion are known at a given point at $t = 0$, the velocity components and coordinates can be found at any later time t. Furthermore, one can also calculate the horizontal range R and maximum height H if v_o and θ_o are known.

- Practice the technique demonstrated in the text to solve relative velocity problems.

SOLUTIONS TO SELECTED END-OF-CHAPTER PROBLEMS

3. Vector $\vec{A}$ is 3.00 units in length and points along the positive x-axis. Vector $\vec{B}$ is 4.00 units in length and points along the negative y-axis. Use graphical methods to find the magnitude and direction of the vectors (a) $\vec{A} + \vec{B}$ and (b) $\vec{A} - \vec{B}$.

Solution

(a) To add vectors graphically, the first vector is drawn in the specified direction and with a length that corresponds to the magnitude of this vector in the chosen scale. For example, if the chosen scale is $1/2$ inch $= 1$ unit, the vector drawn to represent $\vec{A}$ would be 1.50 inches long and point in the positive x-direction. Then, to add $\vec{B}$ to $\vec{A}$ the vector representing $\vec{B}$ is drawn, starting at the tip of $\vec{A}$ and pointing in the specified direction, with a length corresponding to its magnitude on the chosen scale. In this case, the second vector starts at the tip of $\vec{A}$, is drawn in the

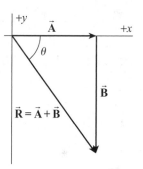

negative y-direction, and has a length of $(4.00 \text{ units})(0.500 \text{ inch}/1 \text{ unit}) = 2.00$ inches. The resultant (vector sum) is the vector drawn from the tail of the first vector to the tip of the last vector as shown in the sketch. The magnitude and direction of the resultant is determined by measuring its length with a ruler (and multiplying by the chosen scale factor) and by measuring its direction with a protractor. In the scale drawing you do, you should find that

$$\vec{R} = \vec{A} + \vec{B} = 5.0 \text{ units at } 53° \text{ below the positive } x\text{-axis} \qquad \Diamond$$

(b) To subtract vector $\vec{B}$ from vector $\vec{A}$, realize that $\vec{A} - \vec{B} = \vec{A} + (-\vec{B})$ and that $-\vec{B}$ is a vector of the same magnitude as $\vec{B}$, but in the opposite direction. The scale drawing you will need to determine $\vec{A} - \vec{B}$ graphically is similar to the sketch at the right. You should find that

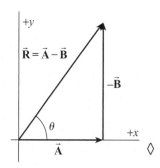

$$\vec{R} = \vec{A} - \vec{B} = 5.0 \text{ units at } 53° \text{ above the positive } x\text{-axis} \qquad \Diamond$$

7. A plane flies from base camp to lake A, a distance of 280 km at a direction of 20.0° north of east. After dropping off supplies, the plane flies to lake B, which is 190 km and 30.0° west of north from lake A. Graphically determine the distance and direction from lake B to the base camp.

Solution

First choose a convenient scale, such as 1 cm = 20 km, and lay out the first two legs of the trip on a scale drawing with the base camp at the origin of the coordinate system. Choose the horizontal axis to be the east-west line and orient the arrow representing the trip to lake A and the arrow representing the trip from lake A to lake B in the specified directions as shown at the right. This process accurately locates the two lakes on your drawing.

The vector $\vec{D}$ drawn from the location of lake B back to the origin represents the displacement of the base camp from lake B. Measure the length of this vector and multiply by your scale factor to determine the distance you are from camp. Measure the angle θ to determine the direction from lake B back to the base camp. Your results should approximately equal the vector given below:

$$\vec{D} = 310 \text{ km at } 57° \text{ south of west} \qquad \Diamond$$

---

17. The eye of a hurricane passes over Grand Bahama Island in a direction 60.0° north of west with a speed of 41.0 km/h. Three hours later, the course of the hurricane suddenly shifts due north, and its speed slows to 25.0 km/h. How far from Grand Bahama is the hurricane 4.50 h after it passes over the island?

Solution

Assume that Grand Bahama Island lies at the origin of the coordinate system shown in the diagram at the right. During the first three hours after passing this island, the hurricane travels a distance

$$A = v_1 \Delta t_1 = (41.0 \text{ km/h})(3.00 \text{ h}) = 123 \text{ km}$$

in a direction 60.0° north of west. In doing so, it moves west a distance

$$A_{\text{west}} = (123 \text{ km})\cos 60.0° = 61.5 \text{ km}$$

and north a distance

$$A_{\text{north}} = (123 \text{ km})\sin 60.0° = 106 \text{ km}$$

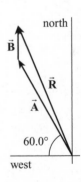

In the next 1.50 h, the hurricane moves due northward an additional distance of

$$B_{north} = v_2 \Delta t_2 = (25.0 \text{ km/h})(1.50 \text{ h}) = 37.5 \text{ km}$$

Thus, 4.50 h after passing Grand Bahama the hurricane is located

$$R_{west} = A_{west} + B_{west} = 61.5 \text{ km} + 0 = 61.5 \text{ km west}$$

and

$$R_{north} = A_{north} + B_{north} = 106 \text{ km} + 37.5 \text{ km} = 144 \text{ km north}$$

Its total distance from the island at this time is

$$R = \sqrt{R_{west}^2 + R_{north}^2} = \sqrt{(61.5 \text{ km})^2 + (144 \text{ km})^2} = 157 \text{ km} \qquad \Diamond$$

19. A commuter airplane starts from an airport and takes the route shown in Figure P3.19. It first flies to city A located 175 km away in a direction 30.0° north of east. Next, it flies for 150 km 20.0° west of north, to city B. Finally, the plane flies 190 km due west, to city C. Find the location of city C relative to the location of the starting point.

Solution

The three successive displacements undergone by the airplane are shown in Figure P3.19 as

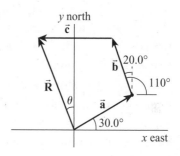

$$\vec{a} = 175 \text{ km at } 30.0° \text{ north of east}$$

$$\vec{b} = 150 \text{ km at } 20.0° \text{ west of north}$$

and

$$\vec{c} = 190 \text{ km directed due west}$$

The displacement of city C relative to the location of the starting point is the vector sum of these three individual displacements. To compute this sum, we start by resolving the displacements $\vec{a}$, $\vec{b}$, and $\vec{c}$ into their x-components (eastward) and y-components (northward). These components are:

$$a_x = +|\vec{a}|\cos 30.0° = +(175 \text{ km})\cos 30.0 = +152 \text{ km}$$

$$a_y = +|\vec{a}|\sin 30.0° = +(175 \text{ km})\sin 30.0 = +87.5 \text{ km}$$

$$b_x = -|\vec{b}|\sin 20.0° = -(150 \text{ km})\sin 20.0° = -51.3 \text{ km}$$

$$b_y = +|\vec{b}|\cos 20.0° = +(150 \text{ km})\cos 20.0° = +141 \text{ km}$$

$$c_x = -190 \text{ km} \quad \text{and} \quad c_y = 0$$

The components of the resultant displacement are found to be

$$R_x = a_x + b_x + c_x = (152 - 51.3 - 190) \text{ km} = -89.7 \text{ km}$$

and

$$R_y = a_y + b_y + c_y = (87.5 + 141 + 0) \text{ km} = +228 \text{ km}$$

The magnitude and direction of this resultant displacement are given by

$$R = |\vec{R}| = \sqrt{R_x^2 + R_y^2} = \sqrt{(-89.7 \text{ km})^2 + (228 \text{ km})^2} = 245 \text{ km}$$

and

$$\theta = \tan^{-1}\left(\frac{|R_x|}{R_y}\right) = \tan^{-1}\left(\frac{89.7 \text{ km}}{228 \text{ km}}\right) = 21.4°$$

or the resultant displacement is

$$\vec{R} = 245 \text{ km at } 21.4° \text{ west of north} \qquad \Diamond$$

23. A student stands at the edge of a cliff and throws a stone horizontally over the edge with a speed of 18.0 m/s. The cliff is 50.0 m above a flat, horizontal beach as shown in Figure P3.23. (a) What are the coordinates of the initial position of the stone? (b) What are the components of the initial velocity? (c) Write the equations for the x- and y-components of the velocity of the stone with time. (d) Write the equations for the position of the stone with time, using the coordinates shown in Figure P3.23. (e) How long after being released does the stone strike the beach below the cliff? (f) With what speed and angle of impact does the stone land?

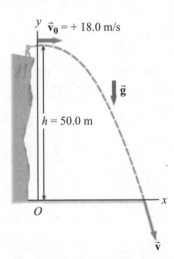

Figure P3.23

Solution

(a) With the origin chosen at point O as shown in Figure P3.23, the coordinates of the original position of the stone are $x_0 = 0$ and $y_0 = +50.0$ m. $\qquad \Diamond$

(b) The components of the initial velocity of the stone are

$$v_{0x} = +18.0 \text{ m/s} \text{ and } v_{0y} = 0 \qquad \Diamond$$

(c) The components of the stone's velocity during its flight are given as functions of time by

$$v_x = v_{0x} + a_x t = 18.0 \text{ m/s} + (0)t \qquad \text{or} \qquad v_x = 18.0 \text{ m/s} \qquad \Diamond$$

and

$$v_y = v_{0y} + a_y t = 0 + (-g)t \qquad \text{or} \qquad v_y = -(9.80 \text{ m/s}^2)t \qquad \Diamond$$

(d) The coordinates of the stone at elapsed time t during its flight are

$$x = x_0 + v_{0x}t + \frac{1}{2}a_x t^2 = 0 + (18.0 \text{ m/s})t + \frac{1}{2}(0)t^2 \quad \text{or} \quad x = (18.0 \text{ m/s})t \qquad \Diamond$$

and

$$y = y_0 + v_{0y}t + \frac{1}{2}a_y t^2 = 50.0 \text{ m} + (0)t + \frac{1}{2}(-g)t^2$$

or

$$y = 50.0 \text{ m} - (4.90 \text{ m/s}^2)t^2 \qquad \Diamond$$

(e) Since $\Delta y = y - y_0 = 0 - 50.0$ m at the instant the stone strikes the beach, we find the time of fall from $\Delta y = v_{0y}t + \frac{1}{2}a_y t^2$ with $v_{0y} = 0$:

$$t = \sqrt{\frac{2(\Delta y)}{a_y}} = \sqrt{\frac{2(-50.0 \text{ m})}{-9.80 \text{ m/s}^2}} = 3.19 \text{ s} \qquad \Diamond$$

(f) At impact, $v_x = v_{0x} = 18.0$ m/s, and the vertical component of velocity is

$$v_y = v_{0y} + a_y t = 0 + (-9.80 \text{ m/s}^2)(3.19 \text{ s}) = -31.3 \text{ m/s}$$

Thus, the speed just before impact is

$$v = \sqrt{v_x^2 + v_y^2} = \sqrt{(18.0 \text{ m/s})^2 + (-31.3 \text{ m/s})^2} = 36.1 \text{ m/s} \qquad \Diamond$$

and the direction of travel is

$$\theta = \tan^{-1}\left(\frac{v_y}{v_x}\right) = \tan^{-1}\left(\frac{-31.3}{18.0}\right) = -60.1° \qquad \Diamond$$

or the velocity just before impact is

$$\vec{v} = 36.1 \text{ m/s at } 60.1° \text{ below the horizontal} \qquad \Diamond$$

27. A tennis player standing 12.6 m from the net hits the ball at 3.00° above the horizontal. To clear the net, the ball must rise at least 0.330 m. If the ball just clears the net at the apex of its trajectory, how fast was the ball moving when it left the racquet?

Solution

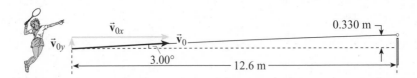

At the apex of the trajectory, $v_y = 0$. Therefore, $v_y = v_{0y} + a_y t$ gives the time to reach the net as

$$t = \frac{v_y - v_{0y}}{a_y} = \frac{0 - v_0 \sin 3.00°}{-g} = \frac{v_0 \sin 3.00°}{g}$$

Since the vertical acceleration is constant, the average velocity in the vertical direction is

$$\left(v_y\right)_{av} = \frac{v_y + v_{0y}}{2}$$

For the time interval from when the ball leaves the racquet until when it reaches the net, this becomes

$$\left(v_y\right)_{av} = \frac{0 + v_0 \sin 3.00°}{2} = \frac{v_0 \sin 3.00°}{2}$$

The vertical displacement during this time interval is

$$\Delta y = \left(v_y\right)_{av} t = \left(\frac{v_0 \sin 3.00°}{2}\right)\left(\frac{v_0 \sin 3.00°}{g}\right) = \frac{v_0^2 \sin^2 3.00°}{2g}$$

If the ball just clears the net, then $\Delta y = 0.330$ m, giving

$$v_0 = \frac{\sqrt{2g(\Delta y)}}{\sin 3.00°} = \frac{\sqrt{2(9.80 \text{ m/s}^2)(0.330 \text{ m})}}{\sin 3.00°} = 48.6 \text{ m/s} \qquad \lozenge$$

Note that it was unnecessary to use the horizontal distance of 12.6 m in this solution.

33. A projectile is launched with an initial speed of 60.0 m/s at an angle of 30.0° above the horizontal. The projectile lands on a hillside 4.00 s later. Neglect air friction. (a) What is the projectile's velocity at the highest point of its trajectory? (b) What is the straight-line distance from where the projectile was launched to where it hits its target?

Solution

(a) At the highest point on the trajectory, the projectile's velocity is horizontal with components of

$$v_y = 0 \text{ and } v_x = v_{0x} = v_0 \cos\theta = (60.0 \text{ m/s})\cos 30.0° = 52.0 \text{ m/s}$$

Thus, at the trajectory's highest point,

$$v = \sqrt{v_x^2 + v_y^2} = 52.0 \text{ m/s}$$

and

$$\vec{v} = 52.0 \text{ m/s directed horizontally} \qquad \Diamond$$

(b) At $t = 4.00 \ s$ after launch, the displacement of the projectile from the launch site has components of

$$\Delta x = v_{0x}t = (v_0 \cos\theta)t = (60.0 \text{ m/s})\cos 30.0°(4.00 \text{ s}) = 208 \text{ m}$$

and

$$\Delta y = v_{0y}t + \tfrac{1}{2}a_y t^2 = (v_0 \sin\theta)t - \tfrac{1}{2}gt^2$$

$$= \left[(60.0 \text{ m/s})\sin 30.0°\right](4.00 \text{ s}) - \frac{1}{2}(9.80 \text{ m/s}^2)(4.00 \text{ s})^2 = 41.6 \text{ m}$$

The straight-line distance from the launch site to the landing spot equals the magnitude of the displacement vector, or

$$d = \sqrt{(\Delta x)^2 + (\Delta y)^2} = \sqrt{(208 \text{ m})^2 + (41.6 \text{ m})^2} = 212 \text{ m} \qquad \Diamond$$

43. A bomber is flying horizontally over level terrain at a speed of 275 m/s relative to the ground and at an altitude of 3.00 km. (a) The bombardier releases one bomb. How far does the bomb travel horizontally between its release and its impact on the ground? Ignore the effects of air resistance. (b) Firing from the people on the ground suddenly incapacitates the bombardier before he can call, "Bombs away!" Consequently, the pilot maintains the plane's original course, altitude, and speed through a storm of flak. Where is the plane relative to the bomb's point of impact when the bomb hits the ground? (c) The plane has a telescopic bombsight set so that the bomb hits the target seen in the sight at the moment of release. At what angle from the vertical was the bombsight set?

Solution

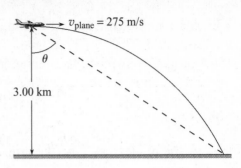

(a) At the instant of its release, the bomb is moving horizontally with the velocity of the plane. Thus, its initial velocity components are $v_{0x} = v_{plane} = 275$ m/s, and $v_{0y} = 0$. With $a_y = -g$, $\Delta y = v_{0y}t + a_y t^2/2$ gives the time required for the bomb to fall to the ground as

$$t = \sqrt{\frac{2(\Delta y)}{a_y}} = \sqrt{\frac{2(-3.00 \times 10^3 \text{ m})}{-9.80 \text{ m/s}^2}} = 24.7 \text{ s}$$

With $a_x = 0$, the horizontal distance the bomb travels during this time is

$$\Delta x = v_{0x}t + a_x t^2/2 = (275 \text{ m/s})(24.7 \text{ s}) + 0 = 6.80 \times 10^3 \text{ m} = 6.80 \text{ km} \qquad \Diamond$$

(b) Since the plane maintains its original course and speed, it and the bomb travel in the same horizontal direction and at the same constant horizontal speeds. As a result, the plane stays directly above the bomb as the bomb drops to the ground. The plane will be 3.00 km directly above the point of impact when the bomb hits the ground. $\qquad \Diamond$

(c) As shown in the above diagram, the line of sight from the bomb's release point to point of impact makes angle θ with the vertical, where

$$\tan\theta = \frac{\text{horizontal distance bomb travels}}{\text{vertical distance bomb falls}} = \frac{\Delta x}{y_0} = \frac{6.80 \text{ km}}{3.00 \text{ km}} = 2.67$$

Thus, the bombsight should be set at $\theta = \tan^{-1}(2.27) = 66.2°$ from the vertical. $\qquad \Diamond$

---

51. A rocket is launched at an angle of 53.0° above the horizontal with an initial speed of 100 m/s. The rocket moves for 3.00 s along its initial line of motion with an acceleration of 30.0 m/s². At this time its engines fail and the rocket proceeds to move as a projectile. Find (a) the maximum altitude reached by the rocket, (b) its total time of flight, and (c) its horizontal range.

Solution

The distance, s, moved in the first 3.00 seconds is given by

$$s = v_0 t + \frac{1}{2}at^2 = (100 \text{ m/s})(3.00 \text{ s}) + \frac{1}{2}(30.0 \text{ m/s}^2)(3.00 \text{ s})^2 = 435 \text{ m}$$

Taking the origin at the point where the rocket was launched, and upward as the positive y-direction, the coordinates of the rocket at the end of powered flight are

$$x_1 = s\cos 53.0° = 262 \text{ m} \quad \text{and} \quad y_1 = s\sin 53.0° = 347 \text{ m}$$

The speed of the rocket at the end of powered flight is

$$v_1 = v_0 + at = 100 \text{ m/s} + (30.0 \text{ m/s}^2)(3.00 \text{ s}) = 190 \text{ m/s}$$

so the initial velocity components for the free-fall phase of the flight are

$$v_{0x} = v_1 \cos 53.0° = 114 \text{ m/s} \qquad \text{and} \qquad v_{0y} = v_1 \sin 53.0° = 152 \text{ m/s}$$

(a) When the rocket is at maximum altitude, $v_y = 0$. The time during which the rocket continues to gain altitude at the beginning of the free-fall phase can be found from $v_y = v_{0y} + a_y t$ as

$$t_{rise} = \frac{0 - v_{0y}}{a_y} = \frac{0 - 152 \text{ m}}{-9.80 \text{ m/s}^2} = 15.5 \text{ s}$$

The vertical displacement occurring during this rise time is

$$\Delta y_{rise} = \left(\frac{v_y + v_{0y}}{2} \right) t_{rise} = \left(\frac{0 + 152 \text{ m/s}}{2} \right)(15.5 \text{ s}) = 1.17 \times 10^3 \text{ m}$$

The maximum altitude reached during the flight is then

$$H = y_1 + \Delta y_{rise} = 347 \text{ m} + 1.17 \times 10^3 \text{ m} = 1.52 \times 10^3 \text{ m} \qquad \Diamond$$

(b) After reaching the top of the arc, the rocket falls 1.52×10^3 m to the ground, starting with zero vertical velocity ($v_{0y} = 0$). The time for this fall is found from $\Delta y = v_{0y} t + \frac{1}{2} a_y t^2$ as

$$t_{fall} = \sqrt{\frac{2(\Delta y)}{a_y}} = \sqrt{\frac{2(-H)}{-g}} = \sqrt{\frac{2(-1.52 \times 10^3 \text{ m})}{-9.80 \text{ m/s}^2}} = 17.6 \text{ s}$$

Therefore, the total time of flight is

$$t = t_{powered} + t_{rise} + t_{fall} = (3.00 + 15.5 + 17.6) \text{ s} = 36.1 \text{ s} \qquad \Diamond$$

(c) The duration of the free-fall phase of the flight is

$$t_2 = t_{rise} + t_{fall} = (15.5 + 17.6) \text{ s} = 33.1 \text{ s}$$

The horizontal displacement occurring during free-fall phase is

$$\Delta x_{free \, fall} = v_{0x} t_2 = (114 \text{ m/s})(33.1 \text{ s}) = 3.78 \times 10^3 \text{ m}$$

and the total horizontal displacement occurring during the full flight (i.e., the horizontal range of the rocket) is

$$R = x_1 + \Delta x_{free \, fall} = 262 \text{ m} + 3.78 \times 10^3 \text{ m} = 4.05 \times 10^3 \text{ m} \qquad \Diamond$$

55. A home run is hit in such a way that the baseball just clears a wall 21 m high, located 130 m from home plate. The ball is hit at an angle of 35° to the horizontal, and air resistance is negligible. Find (a) the initial speed of the ball, (b) the time it takes the ball to reach the wall, and (c) the velocity components and the speed of the ball when it reaches the wall. (Assume the ball is hit at a height of 1.0 m above the ground.)

Solution

(a) Choose a reference frame with the origin at the point where the ball leaves the bat. The x-axis should be horizontal and directed toward the point where the ball crosses the wall. The time required for the ball to reach the wall (i.e. achieve a horizontal displacement of 130 m) is

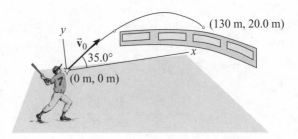

$$t = \frac{\Delta x}{v_{0x}} = \frac{130 \text{ m}}{v_0 \cos 35°} = \frac{159 \text{ m}}{v_0}$$

At this time, the ball must be 21 m above the ground, or 20 m above its launch point ($\Delta y = +20$ m). Therefore, $\Delta y = v_{0y}t + \frac{1}{2}a_y t^2$ becomes

$$20 \text{ m} = \left(\cancel{v_0} \sin 35° \right)\left(\frac{159 \text{ m}}{\cancel{v_0}} \right) + \frac{1}{2}\left(-9.80 \text{ m/s}^2 \right)\left(\frac{159 \text{ m}}{v_0} \right)^2$$

Simplifying and solving for the initial velocity gives

$$v_0 = \sqrt{\frac{\left(-9.80 \text{ m/s}^2 \right)\left(159 \text{ m} \right)^2}{2\left(20 \text{ m} - \left(159 \text{ m} \right)\sin 35° \right)}} = 42 \text{ m/s} \qquad \Diamond$$

(b) From above, the elapsed time when the ball reaches the wall is

$$t = \frac{159 \text{ m}}{v_0} = \frac{159 \text{ m}}{42 \text{ m/s}} = 3.8 \text{ s} \qquad \Diamond$$

(c) At this time, the velocity components of the ball are

$$v_x = v_{0x} = v_0 \cos 35° = \left(42 \text{ m/s} \right)\cos 35° = 34 \text{ m/s}, \qquad \Diamond$$

and

$$v_y = v_{0y} + a_y t = \left(42 \text{ m/s} \right)\sin 35° + \left(-9.80 \text{ m/s}^2 \right)\left(3.8 \text{ s} \right) = -13 \text{ m/s} \qquad \Diamond$$

The speed of the ball as it crosses the wall is $v = \sqrt{v_x^2 + v_y^2} = 37 \text{ m/s}$ $\qquad \Diamond$

61. By throwing a ball at an angle of 45°, a girl can throw the ball a maximum horizontal distance of R on a level field. How far can she throw the same ball vertically upward? Assume that her muscles give the ball the same speed in each case. (Is this assumption valid?)

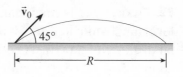

Solution

Throwing the ball at an angle of 45° achieves maximum range for a given initial speed. At this projection angle, the initial velocity components are

$$v_{0x} = v_{0y} = v_0 \sin 45° = v_0/\sqrt{2}$$

We realize that when the ball returns to ground level, it will have $v_y = -v_{0y}$ and make use of $v_y = v_{0y} + a_y t$ to determine the time of flight as

$$t = \frac{v_y - v_{0y}}{a_y} = \frac{-v_{0y} - v_{0y}}{-g} = \frac{2v_{0y}}{g} = \frac{2}{g}\left(\frac{v_0}{\sqrt{2}}\right) = \frac{v_0\sqrt{2}}{g}$$

The maximum horizontal range is then

$$R = v_{0x}t = \left(\frac{v_0}{\sqrt{2}}\right)\left(\frac{v_0\sqrt{2}}{g}\right) = \frac{v_0^2}{g} \tag{1}$$

When the ball is thrown straight upward at speed v_0, the velocity at maximum height is $v_y = 0$ and $v_y^2 = v_{0y}^2 + 2a_y \Delta y$ gives the maximum height reached as

$$(\Delta y)_{max} = \frac{v_y^2 - v_{0y}^2}{2a_y} = \frac{0 - v_0^2}{-2g} = \frac{v_0^2}{2g}$$

Comparing this result to Equation (1) shows that for a given initial speed,

$$(\Delta y)_{max} = \frac{R}{2} \qquad \Diamond$$

If the girl takes a step when she makes the horizontal throw, she can likely give a higher initial speed to that throw than for the vertical throw. $\qquad \Diamond$

72. A dart gun is fired while being held horizontally at a height of 1.00 m above ground level, and while it is at rest relative to the ground. The dart from the gun travels a horizontal distance of 5.00 m. A college student holds the same gun in a horizontal position while sliding down a 45.0° incline at a constant speed of 2.00 m/s. How far will the dart travel if the student fires the gun when it is 1.00 m above the ground?

Solution

When the dart if fired from a stationary gun, the initial velocity components are $v_{0x} = |\vec{v}_{DG}|$ and $v_{0y} = 0$, where $|\vec{v}_{DG}|$ is the speed with which the dart emerges from the gun. In this case, $\Delta y = v_{0y}t + \frac{1}{2}a_yt^2$ gives the time required for the dart to reach the ground as

$$t = \sqrt{\frac{2\Delta y}{a_y}} = \sqrt{\frac{2(-1.00 \text{ m/s})}{-9.80 \text{ m/s}^2}} = 0.452 \text{ s}$$

Thus, the speed with which the dart emerges from the gun is

$$|\vec{v}_{DG}| = v_{0x} = \frac{\Delta x}{t} = \frac{5.00 \text{ m}}{0.452 \text{ s}} = 11.1 \text{ m/s}$$

When the dart is fired horizontally from the moving gun, the initial velocity of the dart relative to the gun, $\vec{v}_{DG}$, is the difference between the initial velocity of the dart relative to Earth, $\vec{v}_{DE}$, and the velocity of the gun relative to Earth, $\vec{v}_{GE}$.

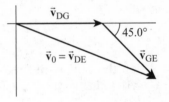

That is,

$$\vec{v}_{DG} = \vec{v}_{DE} - \vec{v}_{GE}$$

The initial velocity of the dart relative to Earth is then $\vec{v}_0 = \vec{v}_{DE} = \vec{v}_{DG} + \vec{v}_{GE}$, with components of

$$v_{0x} = |\vec{v}_{DG}|_x + |\vec{v}_{GE}|_x = 11.1 \text{ m/s} + (2.00 \text{ m/s})\cos 45.0° = 12.5 \text{ m/s}$$

and

$$v_{0y} = |\vec{v}_{DG}|_y + |\vec{v}_{GE}|_y = 0 - (2.00 \text{ m/s})\sin 45.0° = -1.41 \text{ m/s}$$

The vertical speed of the dart when it reaches the ground (1.00 m below its launch point) in this case is given by $v_y^2 = v_{0y}^2 + 2a_y(\Delta y)$ as

$$v_y = -\sqrt{v_{0y}^2 + 2a_y(\Delta y)} = -\sqrt{(-1.41 \text{ m/s})^2 + 2(-9.80 \text{ m/s}^2)(-1.00 \text{ m})} = -4.65 \text{ m/s}$$

The time of flight of the dart in this case is then found from $v_y = v_{0y} + a_yt$ to be

$$t = \frac{v_y - v_{0y}}{a_y} = \frac{-4.65 \text{ m/s} - (-1.41 \text{ m/s})}{-9.80 \text{ m/s}^2} = 0.330 \text{ s}$$

and the horizontal distance the dart travels during this flight is

$$\Delta x = v_{0x}t = (12.5 \text{ m/s})(0.330 \text{ s}) = 4.12 \text{ m} \qquad \lozenge$$

4

Laws of Motion

NOTES FROM SELECTED CHAPTER SECTIONS

4.1 Forces

Equilibrium is the condition under which the net force (vector sum of all forces) acting on an object is zero. An object in equilibrium has a zero acceleration (velocity is constant or equals zero).

Fundamental forces in nature are (listed in order of decreasing strength):

- strong nuclear forces (between subatomic particles)

- electromagnetic forces (between electric charges at rest or in motion)

- weak nuclear forces (accompanying the process of radioactive decay)

- gravitational (attractive forces between objects due to their mass)

Classical physics is concerned with **contact forces** (which are the result of physical contact between two or more objects) and **action-at-a-distance forces** (which act through empty space and do not involve physical contact).

4.2 Newton's First Law

Newton's first law is called the **law of inertia** and states that an object at rest will remain at rest and an object in motion will remain in motion with a constant velocity unless acted on by a net external force.

Mass and **weight** are two different physical quantities. The weight of a body is equal to the force of gravity acting on the body and varies with location in the Earth's gravitational field. Mass is an inherent property of a body and is a measure of the body's inertia (resistance to change in its state of motion). The SI unit of mass is the **kilogram (kg)** and the unit of weight is the **newton (N).**

4.3 Newton's Second Law

Newton's second law, the **law of acceleration,** states that the acceleration of an object is directly proportional to the resultant (or net) force acting on it and inversely proportional to its mass. *The direction of the acceleration is in the direction of the net force.*

4.4 Newton's Third Law

Newton's third law, the **law of action-reaction,** states that when two bodies interact, the force which body "A" exerts on body "B" (the **action force**) is equal in magnitude and opposite in direction to the force which body "B" exerts on body "A" (the **reaction force**). A consequence of the third law is that forces occur in pairs. *Remember that the action force and the reaction force never cancel when Newton's 2nd law is applied to one of the bodies because only one of these forces acts on this body while the other force acts on the other body.*

4.5 Applications of Newton's Laws

Construction of a **free-body diagram** is an important step in the application of Newton's laws of motion to solve problems involving bodies in equilibrium or accelerating under the action of external forces. The diagram should **include a labeled arrow to identify each of the external forces** acting on the body whose motion (or condition of equilibrium) is to be studied. *Forces which are the reactions to external forces must not be included.* When a system consists of more than one body or mass, you must construct a free-body diagram for each mass.

4.6 Forces of Friction

When a body is in motion either on a surface or through a viscous medium such as air or water, there is resistance to the motion because the body interacts with its surroundings. We call such resistance a force of friction. Experiments show that the frictional force arises from the nature of the two surfaces. To a good approximation, both $f_{s,\max}$ (maximum force of static friction) and f_k (force of kinetic friction) are proportional to the normal force at the interface between the two surfaces.

EQUATIONS AND CONCEPTS

A **quantitative measurement of mass** (the term used to measure inertia) can be made by comparing the accelerations that a given force will produce on different bodies. If a given force acting on a body of mass m_1 produces an acceleration a_1 and the same force acting on a body of mass m_2 produces an acceleration a_2 the ratio of the two masses equals the inverse of the ratio of the two accelerations.

$$\frac{m_1}{m_2} = \frac{a_2}{a_1}$$

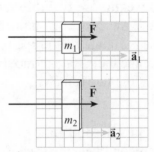

Newton's second law states that the acceleration of an object is proportional to the resultant force acting on it and inversely proportional to its mass.

$$\vec{a} = \frac{\vec{F}}{m}$$

$$\sum \vec{F} = m\vec{a} \qquad (4.1)$$

Component scalar equations are equivalent to the vector form of the equation expressing Newton's second law. *The orientation of the coordinate system can often be chosen so that the object has a nonzero acceleration along only one direction.*

$$\sum F_x = m a_x \qquad (4.2)$$
$$\sum F_y = m a_y$$
$$\sum F_z = m a_z$$

The **SI unit of force** is the newton (N), defined as the force that, when acting on a 1-kg mass, produces an acceleration of 1 m/s². In the U.S. customary system, the unit of force is the pound. *Calculations with Equations (4.1) and (4.2) must be made using a consistent set of units for the quantities force, mass, and acceleration.*

$$1 \text{ N} \equiv 1 \text{ kg} \cdot \text{m/s}^2 \qquad (4.3)$$
$$1 \text{ N} = 0.225 \text{ lb} \qquad (4.4)$$

Newton's law of universal gravitation states that every particle in the universe attracts every other particle with a force that is directly proportional to the product of the masses of the particles and inversely proportional to the square of the distance between them.

$$F_g = G \frac{m_1 m_2}{r^2} \qquad (4.5)$$

The **universal gravitational constant.**

$$G = 6.67 \times 10^{-11} \text{ N} \cdot \text{m}^2 / \text{kg}^2$$

Weight is not an inherent property of a body, but depends on the local value of g and varies with location.

$$w = mg \qquad (4.6)$$

The **acceleration due to gravity,** g, decreases with increasing distance from the center of the Earth.

$$g = G \frac{M_E}{r^2} \qquad (4.8)$$

Newton's third law, states that forces always occur in pairs; the force $\vec{F}_{12}$ exerted by body 1 on body 2 (the action force) is equal in magnitude and opposite in direction to the force $\vec{F}_{21}$ exerted by body 2 on body 1 (the reaction force). *The action and reaction forces always act on different objects; they cannot cancel.*

$$\vec{F}_{12} = -\vec{F}_2$$

$$\vec{F}_{12} = -\vec{F}_{21}$$

Equilibrium is a condition of rest or motion with constant velocity (magnitude and direction). *The vector sum of all forces acting on an object in equilibrium is zero.*

$$\sum \vec{F} = 0 \qquad\qquad (4.9)$$

Two-dimensional equilibrium requires that the sum of all forces in the x and y directions must separately equal zero.

$$\sum F_x = 0 \quad \text{and} \quad \sum F_y = 0 \qquad (4.10)$$

The **force of static friction** acting on two objects whose surfaces are in contact, but not in motion relative to each other, cannot be greater than $\mu_s n$, where n is the magnitude of the normal (perpendicular) force between the two surfaces and μ_s is a dimensionless constant which depends on the nature of the pair of surfaces.

$$f_s \leq \mu_s n \qquad\qquad (4.11)$$

μ_s = coefficient of static friction

A **force of kinetic friction** arises when two surfaces are in relative motion. The force on each body is directed opposite the direction of motion of that body. *The coefficient of kinetic friction μ_k depends on the nature of the two surfaces.* Generally, for a given pair of surfaces, $\mu_k < \mu_s$.

$$f_k = \mu_k n \qquad\qquad (4.12)$$

μ_k = coefficient of kinetic friction

SUGGESTIONS, SKILLS, AND STRATEGIES

For problems involving objects in equilibrium,

1. Make a sketch of the situation described in the problem statement.

2. Draw a free-body diagram for the isolated object under consideration and label all external forces acting on the object. Try to guess the correct direction for each force. If you select a direction that leads to a negative sign in your solution for a force, do not be alarmed; this merely means that the direction of the force is the opposite of what you assumed.

3. Establish a coordinate system and resolve all forces into x- and y-components.

4. Use the equations $\sum F_x = 0$ and $\sum F_y = 0$ (for objects in equilibrium, acceleration equals zero). Remember to keep track of the signs of the various force components.

5. Application of step 4 above leads to a set of equations with several unknowns. Solve the simultaneous equations for the unknowns in terms of the known quantities.

For problems involving the application of Newton's second law:

1. Draw a simple, neat diagram of the system and indicate the forces with arrows. Label each force with a symbol representing the nature of the force.

2. Isolate the object of interest whose motion is being analyzed. Draw a free-body diagram for this object; that is, a diagram showing all external forces acting on the object. **For systems containing more than one object, draw a separate diagram for each object. Do not include forces that the object of interest exerts on other objects.**

3. Establish convenient coordinate axes for each object and find the components of the forces along these axes. It is usually convenient to choose the coordinate system so that one of the axes is parallel to the direction of motion of the object.

4. Apply Newton's second law in component form, ($\sum F_x = ma_x$ and $\sum F_y = ma_y$) for each object under consideration.

5. Solve the component equations for the unknowns. *Remember that you must have as many independent equations as you have unknowns in order to obtain a complete solution.* Often in solving such problems, one must also use the equations of kinematics (motion with constant acceleration) from Chapter 2 to find all the unknowns.

REVIEW CHECKLIST

- State in your own words a description of Newton's laws of motion; recall physical examples of each law, and identify the action-reaction force pairs in a multiple-body interaction problem as specified by Newton's third law.

- Express the normal force in terms of other forces acting on an object, and write out the equation which relates the coefficient of friction, force of friction and normal force between an object and the surface on which it rests or moves.

- Apply Newton's laws of motion to various mechanical systems using the recommended procedure discussed in Section 4.5. Identify all external forces acting on each object of interest, draw separate free-body diagrams for each body of interest in the system, and apply Newton's second law, $\vec{F} = m\vec{a}$, in component form.

- Apply the equations of kinematics (which involve the quantities displacement, velocity, and acceleration) as described in Chapter 2 along with those methods and equations of Chapter 4 (involving mass, force, and acceleration) to the solutions of problems where both the kinematic and dynamic aspects are present.

- Be familiar with methods for solving two or more linear equations simultaneously for the unknown quantities. Recall that you must have as many independent equations as you have unknowns.

SOLUTIONS TO SELECTED END-OF-CHAPTER PROBLEMS

9. A Chinook salmon has a maximum underwater speed of 3.0 m/s and can jump out of water vertically with a speed of 6.0 m/s. A record salmon has a length of 1.5 m and a mass of 61 kg. When swimming upward at constant speed, and neglecting buoyancy, the fish experiences three forces: an upward force F exerted by the tail fin, the downward drag force of the water, and the downward force of gravity. As the fish leaves the surface of the water, however, it experiences a net upward force causing it to accelerate from 3.0 m/s to 6.0 m/s. Assuming that the drag force disappears as soon as the head of the fish breaks the surface and that F is exerted until $\frac{2}{3}$ of the fish's length has left the water, determine the magnitude of F.

Solution

As the salmon approaches the surface, three forces act on it as shown in the sketch at the right. However, we assume that the drag force ceases to exist as soon as the fish's nose reaches the surface. This leaves a net upward force of

$$F_{net} = \Sigma F_y = F - mg$$

acting on the fish until it has undergone an additional upward displacement of $\Delta y = \frac{2}{3}(length) = \frac{2}{3}(1.5 \text{ m}) = 1.0 \text{ m}$, when the thrust, F, ceases to act. While the fish undergoes this displacement, its upward acceleration is given by

$$a_y = \frac{F_{net}}{m} = \frac{F}{m} - g$$

or the magnitude of the thrust of the tail fin is $F = m(g + a_y)$.

If the fish's velocity increases from 3.0 m/s to 6.0 m/s during this interval, the vertical acceleration is found from $v_y^2 = v_{0y}^2 + 2a_y(\Delta y)$ as

$$a_y = \frac{v_y^2 - v_{0y}^2}{2(\Delta y)} = \frac{(6.0 \text{ m/s})^2 - (3.0 \text{ m/s})^2}{2(1.0 \text{ m})} = 13.5 \text{ m/s}^2$$

so the thrust due to the tail fin must have been

$$F = m(g + a_y) = (61 \text{ kg})(9.80 \text{ m/s}^2 + 13.5 \text{ m/s}^2) = 1.4 \times 10^3 \text{ N} \qquad \Diamond$$

16. The force exerted by the wind on the sails of a sailboat is 390 N north. The water exerts a force of 180 N east. If the boat (including its crew) has a mass of 270 kg, what are the magnitude and direction of its acceleration?

Solution

We choose east to be the positive x-direction and north the positive y-direction. The horizontal forces acting on the 270-kg boat (including crew) are shown at the right.

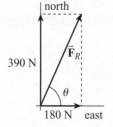

One method of finding the desired acceleration is to first compute the resultant force $\vec{F}_R$:

$$F_R = \sqrt{\left(\Sigma F_x\right)^2 + \left(\Sigma F_y\right)^2} = \sqrt{\left(180 \text{ N}\right)^2 + \left(390 \text{ N}\right)^2} = 430 \text{ N}$$

$$\theta = \tan^{-1}\left(\frac{\Sigma F_y}{\Sigma F_x}\right) = \tan^{-1}\left(\frac{390 \text{ N}}{180 \text{ N}}\right) = 65.2°$$

or

$$\vec{F}_R = 430 \text{ N at } 65.2° \text{ north of east}$$

Then, the resultant acceleration of the boat is

$$\vec{a}_R = \frac{\vec{F}_R}{m} = \frac{430 \text{ N}}{270 \text{ kg}} \text{ at } 65.2° \text{ north of east} = 1.59 \text{ m/s}^2 \text{ at } 65.2° \text{ north of east} \qquad \Diamond$$

Alternatively, we may first compute the components of the acceleration as

$$a_x = \frac{\Sigma F_x}{m} = \frac{180 \text{ N}}{270 \text{ kg}} = 0.667 \text{ m/s}^2 \qquad \text{and} \qquad a_y = \frac{\Sigma F_y}{m} = \frac{390 \text{ N}}{270 \text{ kg}} = 1.44 \text{ m/s}^2$$

The resultant acceleration is then

$$a_R = \sqrt{\left(a_x\right)^2 + \left(a_y\right)^2} = \sqrt{\left(0.667 \text{ m/s}^2\right)^2 + \left(1.44 \text{ m/s}^2\right)^2} = 1.59 \text{ m/s}^2$$

$$\theta = \tan^{-1}\left(\frac{a_y}{a_x}\right) = \tan^{-1}\left(\frac{1.44 \text{ m/s}^2}{0.667 \text{ m/s}^2}\right) = 65.2°$$

or

$$\vec{a}_R = 1.59 \text{ m/s}^2 \text{ at } 65.2° \text{ north of east} \qquad \Diamond$$

23. The distance between two telephone poles is 50.0 m. When a 1.00-kg bird lands on the telephone wire midway between the poles, the wire sags 0.200 m. Draw a free-body diagram of the bird. How much tension does the bird produce in the wire? Ignore the weight of the wire.

Solution

The needed free-body diagram is given below:

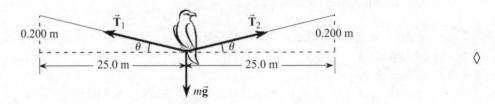

Observe from the dimensions shown in the above diagram, the angle the wire makes with the horizontal is the same on each side of the bird when the bird is located at the center of the wire. This angle is

$$\theta = \tan^{-1}\left(\frac{0.200 \text{ m}}{25.0 \text{ m}}\right) = 0.458°$$

While the bird is resting on the wire, it has zero acceleration. Hence, $a_x = a_y = 0$. Considering the horizontal direction, Newton's second law gives

$$\Sigma F_x = 0 \quad \Rightarrow \quad T_2 \cos\theta - T_1 \cos\theta = 0 \quad \text{or} \quad T_2 = T_1 = T$$

where T is the common value of the tension in the wire on the two sides of the bird.

Now, considering the vertical direction, Newton's second law yields

$$\Sigma F_y = 0 \quad \Rightarrow \quad T_2 \sin\theta + T_1 \sin\theta - mg = 0$$

or since $T_2 = T_1 = T$,

$$2T \sin\theta = mg$$

which gives

$$T = \frac{mg}{2 \sin\theta} = \frac{(1.00 \text{ kg})(9.80 \text{ m/s}^2)}{2 \sin(0.458°)} = 613 \text{ N}$$

29. Assume the three blocks portrayed in Figure P4.29 move on a frictionless surface and a 42-N force acts as shown on the 3.0-kg block. Determine (a) the acceleration given this system, (b) the tension in the cord connecting the 3.0-kg and the 1.0-kg blocks, and (c) the force exerted by the 1.0-kg block on the 2.0-kg block.

Solution

(a) If we consider our system of interest to consist of all three blocks plus the connecting rope, the total mass of the system is

$$m_{total} = (1.0 + 2.0 + 3.0) \text{ kg} = 6.0 \text{ kg}$$

The only horizontal external force acting on this system is the applied 42-N force directed toward the right (chosen as the positive x-direction). Thus, Newton's second law gives

$$a_x = \frac{\Sigma F_x}{m_{total}} = \frac{42 \text{ N}}{6.0 \text{ kg}} = 7.0 \text{ m/s}^2 \text{ toward the right} \qquad \Diamond$$

(b) To determine the tension in the rope, define a new system consisting of only the 3.0-kg block. The external horizontal forces acting on this system are the applied 42-N force and the tension in the rope as shown in the free-body diagram at the right. Then,

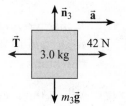

$$\Sigma F_x = ma_x \implies 42 \text{ N} - T = m_3 a_x$$

or

$$T = 42 \text{ N} - m_3 a_x = 42 \text{ N} - (3.0 \text{ kg})(7.0 \text{ m/s}^2) = 21 \text{ N} \qquad \Diamond$$

(c) If we now consider a new system consisting of only the 2.0-kg block, the horizontal force exerted on it by the 1.0-kg block will be an external force as shown in the free-body diagram at the right. Applying Newton's second law to this system gives

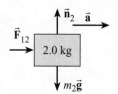

$$\Sigma F_x = ma_x \implies F_{12} = m_2 a_x = (2.0 \text{ kg})(7.0 \text{ m/s}^2) = 14 \text{ N} \qquad \Diamond$$

35. (a) An elevator of mass m moving upward has two forces acting on it, the upward force of tension in the cable and the downward force due to gravity. When the elevator is accelerating upward, which is greater, T or w? (b) When the elevator is moving at a constant velocity upward, which is greater, T or w? (c) When the elevator is moving upward, but the acceleration is downward, which is greater, T or w? (d) Let the elevator have a mass of 1 500 kg and an upward acceleration of 2.50 m/s^2. Find T. Is your answer consistent with the answer to part (a)? (e) The elevator of part (d) now moves with a constant velocity upward of 10 m/s. Find T. Is your answer consistent with your answer of part (b)? (f) Having initially moved upward with a constant velocity, the elevator begins to accelerate downward at 1.50 m/s^2. Find T. Is your answer consistent with your answer to part (c)?

Solution

In considering the motion of an object, it is very important to keep in mind that *the acceleration of the object is always in the direction of the net force* acting on it.

(a) When the elevator is accelerating upward, there must be a net upward force acting on it. Thus, the tension in the cable must exceed the weight of the elevator. ◊

(b) While moving at constant velocity, the elevator has zero acceleration. Hence, the net force acting on it must be zero, or the tension in the cable equals the weight of the elevator. ◊

(c) When the acceleration is directed downward, the net force must be in the downward direction, regardless of the direction of the elevator's velocity. Therefore, the weight w is greater than the tension T in this case. ◊

From Newton's second law: $\Sigma F_y = T - w = ma_y$ or $T = w + ma_y$.

(d) When $a_y = +2.50$ m/s^2, the tension in the cable is

$$T = w + (1\,500 \text{ kg})(+2.50 \text{ m/s}^2)$$
$$= w + 3.75 \times 10^3 \text{ N} > w; \text{ consistent with (a)}$$ ◊

or, since $w = mg = (1\,500 \text{ kg})(9.80 \text{ m/s}^2) = 1.47 \times 10^4 \text{ N}$,

$$T = 1.47 \times 10^4 \text{ N} + 3.75 \times 10^3 \text{ N} = 1.85 \times 10^4 \text{ N}$$ ◊

(e) When v is constant, $a_y = 0$ and $T = w + 0 = w = 1.47 \times 10^4$ N; consistent with (b). ◊

(f) When $a_y = -1.50$ m/s^2,

$$T = w + (1\,500 \text{ kg})(-1.50 \text{ m/s}^2) = 1.47 \times 10^4 \text{ N} - 2.25 \times 10^3 \text{ N}$$

$$T = 1.25 \times 10^4 \text{ N} < w; \text{ consistent with (c)}$$ ◊

43. Consider a large truck carrying a heavy load, such as steel beams. A significant hazard for the driver is that the load may slide forward, crushing the cab, if the truck stops suddenly in an accident or even in braking. Assume, for example, that a 10 000-kg load sits on the flat bed of a 20 000-kg truck moving at 12.0 m/s. Assume the load is not tied down to the truck, but has a coefficient of static friction of 0.500 with the truck bed. (a) Calculate the minimum stopping distance for which the load will not slide forward relative to the truck. (b) Is any piece of data unnecessary for the solution?

Solution

(a) The only horizontal force that can act on the load is the friction force exerted by the bed of the truck. If the load is not slipping, this will be a static friction force, f_s. When the truck slows for any reason, the load tends to slide forward on the bed, and the friction force will be directed toward the rear to oppose this motion.

The only vertical forces acting on the load are the upward normal force and the downward force of gravity. Since the load has no vertical acceleration, Newton's second law says that

$$\Sigma F_y = n - m_{\text{load}}g = m_{\text{load}}a_y = 0 \quad \text{or} \quad n = m_{\text{load}}g$$

Thus, the maximum friction force the bed can exert on the load has magnitude

$$\left(f_s\right)_{\text{max}} = \mu_s n = \mu_s m_{\text{load}}g$$

Now, applying Newton's second law in the horizontal direction when the truck is slowing, at the maximum rate without causing the load to slip, gives

$$\Sigma F_x = -\left(f_s\right)_{\text{max}} = m_{\text{load}}\left(a_x\right)_{\text{max}} \quad \text{or} \quad \left(a_x\right)_{\text{max}} = -\frac{\left(f_s\right)_{\text{max}}}{m_{\text{load}}} = -\frac{\mu_s m_{\text{load}}g}{m_{\text{load}}} = -\mu_s g$$

The minimum distance the truck can safely stop in is given by $v_x^2 = v_{0x}^2 + 2a_x(\Delta x)$ as

$$\left(\Delta x\right)_{\text{min}} = \frac{v_x^2 - v_{0x}^2}{2\left(a_x\right)_{\text{max}}} = \frac{0 - v_{0x}^2}{-2\mu_s g} = \frac{v_{0x}^2}{2\mu_s g} = \frac{\left(12.0 \text{ m/s}\right)^2}{2\left(0.500\right)\left(9.80 \text{ m/s}^2\right)} = 14.7 \text{ m} \qquad \Diamond$$

(b) Observe that neither the mass of the truck nor the mass of the load was needed in the solution of this problem. $\qquad \Diamond$

49. An object falling under the pull of gravity is acted upon by a frictional force of air resistance. The magnitude of this force is approximately proportional to the speed of the object, which can be written as $f = bv$. Assume that $b = 15$ kg/s and $m = 50$ kg. (a) What is the terminal speed the object reaches while falling? (b) Does your answer to part (a) depend on the initial speed of the object? Explain.

Solution

(a)   A free-body diagram of the falling body is given at the right. Taking downward as positive, Newton's second law yields

$$\Sigma F_y = mg - bv = ma_y$$

or the downward acceleration of the falling body will be

$$a_y = \frac{mg - bv}{m} = g - \left(\frac{b}{m}\right)v$$

When the falling body achieves a constant downward speed (the terminal speed), the acceleration is $a_y = 0$, or

$$g - \left(\frac{b}{m}\right)v_{terminal} = 0$$

and the terminal velocity is

$$v_{terminal} = \frac{g}{(b/m)} = \frac{mg}{b} = \frac{(50 \text{ kg})(9.8 \text{ m/s}^2)}{15 \text{ kg/s}} = 33 \text{ m/s} \qquad \Diamond$$

(b)   The magnitude of the terminal speed does not depend on the initial speed of the falling body. However, the manner in which the body approaches its terminal speed does depend on the initial speed. If $v_0 < v_{terminal}$ (as when a body starts falling from rest), then $f < mg$ and the object has a downward acceleration, picking up speed until it reaches a maximum value of $v = v_{terminal}$. On the other hand, if $v_0 > v_{terminal}$ (as might be the case of a skydiver just after opening the parachute), then $f > mg$ and the object has an upward acceleration, slowing until it reaches a minimum value of $v = v_{terminal}$. Finally, if $v_0 = v_{terminal}$, then $f = mg$ and the object has zero acceleration, meaning that it continues to fall at a constant speed of $v_0 = v_{terminal}$. $\qquad \Diamond$

55. The person in Figure P4.55 weighs 170 lb. Each crutch makes an angle of 22.0° with the vertical (as seen from the front). Half of the person's weight is supported by the crutches, the other half by the vertical forces exerted by the ground on his feet. Assuming that he is at rest and that the force exerted by the ground on the crutches acts along the crutches, determine (a) the smallest possible coefficient of friction between crutches and ground and (b) the magnitude of the compression force supported by each crutch.

Solution

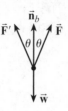

The forces acting on the person's body are shown in the diagram at the right. $\vec{F}$ and $\vec{F}'$ are exerted by the two crutches, $\vec{n}_b$ is a normal force exerted by the ground as it directly supports half the person's weight ($n_b = w/2$), and $\vec{w}$ is the person's weight. The body is in equilibrium, so $a_x = a_y = 0$. Thus,

$$\Sigma F_x = 0 \implies F\sin\theta - F'\sin\theta = 0 \quad \text{or} \quad F' = F$$

Also

$$\Sigma F_y = 0 \implies F\cos\theta + F'\cos\theta + n_b - w = 0$$

With $F' = F$ and $n_b = w/2$, this reduces to

$$2F\cos\theta + w/2 - w = 0 \quad \text{or} \quad F = \frac{w}{4\cos\theta}$$

Consider the free-body diagram of a crutch tip given at the right. Here $\vec{F}$ is the reaction force exerted on the crutch by the body and transmitted as a compression force to the tip. $\vec{n}$ and $\vec{f}_s$ are the normal force and static friction force exerted on the tip by the ground.

$$\Sigma F_y = 0 \implies n = F\cos\theta = \left(\frac{w}{4\cos\theta}\right)\cos\theta = \frac{w}{4}$$

$$\Sigma F_x = 0 \implies f_s = F\sin\theta = \left(\frac{w}{4\cos\theta}\right)\sin\theta = \frac{w}{4}\tan\theta$$

If the coefficient of friction between the crutch tip and the ground is the minimum necessary to prevent slipping, then $f_s = (f_s)_{max} = \mu_s n$ and

$$\mu_s = \frac{f_s}{n} = \frac{(w/4)\tan\theta}{w/4} = \tan\theta$$

(a) With the crutch at angle $\theta = 22.0°$ from the vertical, the minimal coefficient of friction to prevent slipping is $\mu_s = \tan 22.0° = 0.404$. ◊

(b) With $w = 170$ lb and $\theta = 22.0°$, the compression force supported by the crutch is

$$F = \frac{w}{4\cos\theta} = \frac{170 \text{ lb}}{4\cos(22.0°)} = 45.8 \text{ lb}$$ ◊

61. A boy coasts down a hill on a sled, reaching a level surface at the bottom with a speed of 7.0 m/s. If the coefficient of friction between the sled's runners and the snow is 0.050 and the boy and sled together weigh 600 N, how far does the sled travel on the level surface before coming to rest?

Solution

The forces acting on the boy and sled are shown in the sketch. Applying Newton's second law gives

$$\Sigma F_y = ma_y \implies n - mg = 0 \quad \text{or} \quad n = mg$$

Thus,

$$f_k = \mu_k n = \mu_k mg$$

$$\Sigma F_x = ma_x \implies -\mu_k mg = ma_x \quad \text{or} \quad a_x = -\mu_k g$$

Then, $v_x^2 = v_{0x}^2 + 2a_x \Delta x$ with $v_x = 0$ gives the distance the sled with initial speed v_{0x} will coast before stopping. This yields

$$\Delta x = \frac{0 - v_{0x}^2}{2a_x} = \frac{-v_{0x}^2}{2(-\mu_k g)} = \frac{v_{0x}^2}{2\mu_k g}$$

Therefore, if $v_{0x} = 7.0$ m/s and $\mu_k = 0.050$, the coasting distance on the level ground is

$$\Delta x = \frac{(7.0 \text{ m/s})^2}{2(0.050)(9.80 \text{ m/s}^2)} = 50 \text{ m} \qquad \lozenge$$

Note that the numeric value of the weight of the child and sled was never used in this solution. This means that any coaster starting with the given initial speed will go the same distance before stopping provided the coefficient of kinetic friction between runners and snow remains unchanged.

65. A frictionless plane is 10.0 m long and inclined at 35.0°. A sled starts at the bottom with an initial speed of 5.00 m/s up the incline. When the sled reaches the point at which it momentarily stops, a second sled is released from the top of the incline with an initial speed v_i. Both sleds reach the bottom of the incline at the same moment. (a) Determine the distance that the first sled traveled up the incline. (b) Determine the initial speed of the second sled.

Solution

First, consider the free-body diagram given at the right of either sled on the frictionless slope of inclination $\theta = 35.0°$. The acceleration of the sled will be directed down the incline (chosen as the +x-direction) and has magnitude of

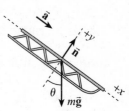

$$a_x = \frac{\Sigma F_x}{m} = \frac{mg\sin\theta}{m} = g\sin\theta$$

(a) If the first sled starts up the incline with speed $v_{0x} = -5.00\,\text{m/s}$ at the bottom, the distance it travels up the incline before stopping is

$$|\Delta x| = \frac{|v_x^2 - v_{0x}^2|}{2a_x} = \frac{|0 - (-5.00\ \text{m/s})^2|}{2\left[(9.80\ \text{m/s}^2)\sin 35.0°\right]} = 2.22\ \text{m} \qquad\qquad \Diamond$$

(b) After stopping momentarily ($v_{0x} = 0$), the time for the first sled to go 2.22 m back down the incline is given by $\Delta x = v_{0x}t + \frac{1}{2}a_x t^2$ as

$$t = \sqrt{\frac{2\Delta x}{a_x}} = \sqrt{\frac{2(2.22\ \text{m})}{(9.80\ \text{m/s}^2)\sin 35.0°}} = 0.890\ \text{s}$$

In this same time, the second sled (with $v_{0x} = +v_i$ and $a_x = g\sin\theta$) must travel the full length of the incline ($\Delta x = 10.0$ m). The required initial speed is given by $\Delta x = v_{0x}t + \frac{1}{2}a_x t^2$ as

$$v_i = \frac{\Delta x}{t} - \frac{1}{2}a_x t = \frac{10.0\ \text{m/s}}{0.890\ \text{s}} - \frac{1}{2}\left[(9.80\ \text{m/s}^2)\sin 35.0°\right](0.890\ \text{s})$$

or

$$v_i = 8.74\ \text{m/s} \qquad\qquad \Diamond$$

73. A van accelerates down a hill (Fig. P4.73), going from rest to 30.0 m/s in 6.00 s. During the acceleration, a toy $(m = 0.100 \text{ kg})$ hangs by a string from the van's ceiling. The acceleration is such that the string remains perpendicular to the ceiling. Determine (a) the angle θ and (b) the tension in the string.

Solution

(a)   The sketch at the right gives a free-body diagram of the van, with the positive x-direction being down the incline. Applying Newton's second law for the x-direction gives

$$\Sigma F_x = mg\sin\theta = ma_x \qquad \text{or} \qquad a_x = g\sin\theta$$

Since the van goes from rest to a velocity of 30.0 m/s down the incline in 6.00 s, its acceleration is given by $v_x = v_{0x} + a_x t$ as

$$a_x = \frac{v_x - v_{0x}}{t} = \frac{30.0 \text{ m/s} - 0}{6.00 \text{ s}} = 5.00 \text{ m/s}^2$$

The angle of the incline must be

$$\theta = \sin^{-1}\left(\frac{a_x}{g}\right) = \sin^{-1}\left(\frac{5.00 \text{ m/s}^2}{9.80 \text{ m/s}^2}\right) = 30.7° \qquad \Diamond$$

(b)   Now consider a free-body diagram of the suspended toy. Since the string remains perpendicular to the ceiling of the van, it must lie along the y-axis in the sketch above, or it is at angle θ from the vertical.

The acceleration of the van and its contents is directed down the incline, or in the positive x-direction. Thus, $a_y = 0$ and Newton's second law gives

$$\Sigma F_y = T - mg\cos\theta = 0$$

or

$$T = mg\cos\theta = (0.100 \text{ kg})(9.80 \text{ m/s}^2)\cos 30.7° = 0.843 \text{ N} \qquad \Diamond$$

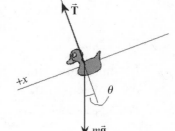

77. The board sandwiched between two other boards in Figure P4.77 weighs 95.5 N. If the coefficient of friction between the boards is 0.663, what must be the magnitude of the compression forces (assumed to be horizontal) acting on both sides of the center board to keep it from slipping?

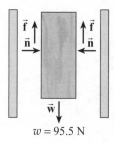

$w = 95.5$ N

Figure P4.77 (modified)

Solution

Since the board is in equilibrium, $\Sigma F_x = 0$ and we see that the normal forces must be the same on both sides of the board. Also, if the minimum normal forces (compression forces) are being applied, the board is on the verge of slipping and the friction force on each side is

$$f = \left(f_s\right)_{max} = \mu_s n$$

The board is also in equilibrium in the vertical direction, so

$$\Sigma F_y = 2f - w = 0 \qquad \text{or} \qquad f = \frac{w}{2}$$

The minimum compression force needed is then

$$n = \frac{f}{\mu_s} = \frac{w}{2\mu_s} = \frac{95.5\text{ N}}{2(0.663)} = 72.0\text{ N} \qquad \Diamond$$

5

Energy

NOTES FROM SELECTED CHAPTER SECTIONS

5.1 Work

In order for work to be accomplished, an object must undergo a displacement; the force associated with the work must have a component parallel to the direction of the displacement. Work is a scalar quantity and can be either positive or negative (positive when the component $F\cos\theta$ is in the same direction as the displacement). The SI unit of work is the newton-meter $(\text{N} \cdot \text{m})$ or joule (J).

5.2 Kinetic Energy and the Work-Energy Theorem

Any object that has mass m and speed v has **kinetic energy.** Kinetic energy is a scalar quantity and has the same units as work; kinetic energy of an object will change only if net work is done on the object by external forces. The relationship between work and change in kinetic energy is stated in the work-energy theorem.

A force is **conservative** if the work it does on an object moving between two points is independent of the path the object takes between the points. The work done on an object by a conservative force depends only on the initial and final positions of the object. The gravitational force is an example of a conservative force.

A force is **nonconservative** if the work it does on an object moving between two points depends on the path taken. Kinetic friction is an example of a nonconservative force.

5.3 Gravitational Potential Energy

The work done on an object by the force of gravity is equal to the object's initial potential energy minus its final potential energy ($W_g = -\Delta PE_g$). The gravitational potential energy associated with an object depends only on the object's weight and its vertical height above the surface of the Earth. If the height above the surface increases, the potential energy will also increase, but the work done by the gravitational force will be negative. (In this case the direction of the displacement is opposite the direction of the gravitational force.) **In working problems involving gravitational potential energy, it is necessary to choose an arbitrary reference level (or location) at which the potential energy is taken to be zero.**

5.4 Spring Potential Energy

Elastic potential energy is the energy associated with a spring that is compressed from or stretched beyond its equilibrium position. The spring force is a conservative force and has a magnitude that is proportional to the displacement of the spring from the equilibrium position.

5.5 Systems and Energy Conservation

The **change in the kinetic energy** of a physical system equals the sum of the work done by conservative forces and the work done by nonconservative forces. The sum of the kinetic energy plus the potential energy is called the **total mechanical energy.** Since the work done by conservative forces equals the negative of the change in potential energy ($W_c = -\Delta PE$), the work done by nonconservative forces equals the change in the total mechanical energy of the system. *The mechanical energy of a system remains constant if only conservative forces do work on the system.*

5.6 Power

Power delivered to an object is defined as the rate at which energy is transferred to the object or the rate at which work is being done on the object. The average power delivered to an object during a time interval can be expressed as the product of the average speed during the time interval and the component of the force in the direction of the velocity.

5.7 Work Done by a Varying Force

If the value of a non-constant force is known as a function of position, the work done by the varying force during a displacement can be determined by calculating the area under the force vs. displacement curve.

EQUATIONS AND CONCEPTS

The **work done by a constant force $\vec{F}$,** (constant in both magnitude and direction) is defined to be the product of the component of the force in the direction of the displacement and the magnitude of the displacement.

$$W \equiv (F\cos\theta)\Delta x \qquad (5.2)$$

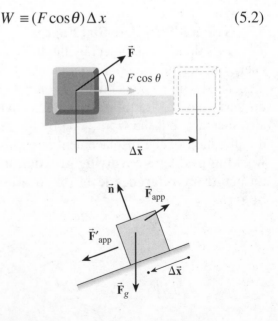

Work can be positive, negative, or zero, depending on the value of θ, the angle between the direction of the force and the direction of the displacement. The figure at the right shows an object acted on by four forces while being displaced down a smooth incline. Consider the done by each of the four forces as illustrated below *when the direction of the x-axis is taken to be along the direction of the incline.*

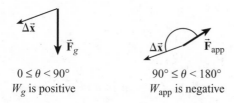
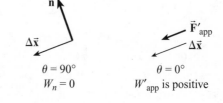

$0 \le \theta < 90°$
W_g is positive

$90° \le \theta < 180°$
W_{app} is negative

$\theta = 90°$
$W_n = 0$

$\theta = 0°$
W'_{app} is positive

Work is a scalar quantity and the SI unit of work is the newton-meter or joule.

$$1 \text{ N} \cdot \text{m} = 1 \text{ J}$$

$$\text{joule (J)} = \text{newton} \cdot \text{meter} = \text{kg} \cdot \text{m}^2/\text{s}^2$$

Kinetic energy, defined by Equation (5.5), is a scalar quantity and is the energy associated with the motion of a mass *m*. Kinetic energy has the same units as work.

$$KE \equiv \tfrac{1}{2}mv^2 \qquad (5.5)$$

The **work-energy theorem** states that the net work done on a particle equals the change in kinetic energy of the particle. *The work-energy theorem is valid for a particle or for a system that can be modeled as a particle.*

$$W_{net} = \tfrac{1}{2}mv^2 - \tfrac{1}{2}mv_0^2 \qquad (5.4)$$

The **net work** done on an object is the sum of the work done by the nonconservative forces and the conservative forces.

$$W_{nc} + W_c = \Delta KE \qquad (5.7)$$

Gravitational potential energy of a mass-Earth system near the surface of the Earth is defined by Equation (5.10). *The quantity y in Equation (5.10) is the vertical position of the mass relative to the Earth (or other arbitrarily chosen reference level where $PE_g = 0$).*

$$PE \equiv mgy \qquad (5.10)$$

The **work done by the force of gravity** can be expressed in terms of initial and final values of the *y*-coordinates. *The units of energy (kinetic and potential) are the same as the units of work. The difference in potential energy between two points is independent of the location of the origin.*

$$W_g = -\left(PE_f - PE_i\right)$$
$$= -\left(mgy_f - mgy_i\right) \qquad (5.11)$$

Hook's law expresses the force exerted by a spring that is stretched or compressed from the equilibrium position. *The negative sign signifies that the force is always directed opposite the displacement from equilibrium.*

$$F_s = -kx \qquad (5.15)$$

k = spring constant in N/m.

Each spring has a characteristic value of k.

Elastic potential energy is associated with a mass-spring system that has been stretched or compressed from the equilibrium position. *The elastic potential energy of a deformed spring is always positive.*

$$PE_s \equiv \tfrac{1}{2}kx^2 \qquad (5.16)$$

x = displacement from equilibrium.

The **work done by all nonconservative forces** equals the change in the total mechanical energy of a system (change in kinetic energy plus changes in the gravitational and elastic potential energies).

$$W_{nc} = (KE_f - KE_i)$$
$$+ (PE_{gf} - PE_{gi}) + (PE_{sf} - PE_{si})$$
$$(5.17)$$

or

$$W_{nc} = \Delta KE + \Delta PE_g + \Delta PE_s$$

The law of conservation of mechanical energy holds when only conservative forces act on a system; the total mechanical energy of the system remains constant.

$$(KE + PE_g + PE_s)_i = (KE + PE_g + PE_s)_f$$
$$(5.18)$$

The **average power** supplied by a force is the ratio of the work done by the force to the time interval over which the force acts. The average power can also be expressed in terms of the force and the average speed of the object on which the force acts. *In Equation (5.23), F is the component of the force along the direction of the velocity.*

$$\overline{\mathcal{P}} \equiv \frac{W}{\Delta t} \qquad (5.22)$$

$$\overline{\mathcal{P}} = F\overline{v} \qquad (5.23)$$

The **SI unit of power** is the watt; in the U.S. customary system, the unit of power is the horsepower.

$$1\ \text{W} = 1\ \text{J/s} = 1\ \text{kg} \cdot \text{m}^2/\text{s}^3 \qquad (5.24)$$

$$1\ \text{hp} = 550\ \frac{\text{ft} \cdot \text{lb}}{\text{s}} = 746\ \text{W} \qquad (5.25)$$

Work done by a variable force is equal to the area under the force vs. displacement curve. The figure at right illustrates the case of a force acting along the x-axis.

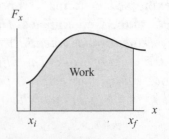

SUGGESTIONS, SKILLS, AND STRATEGIES

Choosing a zero level for potential energy

In working problems involving gravitational potential energy, it is always necessary to choose a location at which to set the gravitational potential energy equal to zero. This choice is completely arbitrary because the important quantity is the difference in potential energy, and that difference is independent of the location of zero. It is often convenient, but not essential, to choose the surface of the Earth as the reference position for zero potential energy. In most cases, the statement of the problem suggests a convenient level to use.

Conservation of energy

Take the following steps in applying the principle of conservation of energy:

1. Define your system, which may consist of more than one object.

2. Select a reference level for the zero point of gravitational potential energy. This level must not be changed during the solution of a specific problem.

3. Determine whether or not nonconservative forces are present.

4. If mechanical energy is conserved (i.e., if only conservative forces are present), you should use Equation (5.18) $(KE + PE_g + PE_s)_i = (KE + PE_g + PE_s)_f$. Usually (except when a quantity is equal to zero) it is better to solve for the unknown algebraically (using symbols) before substituting numerical values.

5. If nonconservative forces such as friction are present (and thus mechanical energy is not conserved), first write expressions for the total initial and total final mechanical energies. In this case, the change in the two total energies is equal to the work done by the nonconservative force(s).

REVIEW CHECKLIST

- Define the work done by a constant force and work done by a force that varies with position. (Recognize that the work done by a force can be positive, negative, or zero; describe at least one example of each case).

- Understand that the work done by a conservative force in moving a body between any two points is independent of the path taken. Nonconservative forces are those for which the work done on a particle moving between two points depends on the path. Account for nonconservative forces acting on a system using the work-energy theorem. In this case, the work done by all nonconservative forces equals the change in total mechanical energy of the system.

- Relate the work done by the net force on an object to the **change in kinetic energy.** The relation $W_{net} = \Delta KE = KE_f - KE_i$ is called the work-energy theorem, and it is valid whether or not the (resultant) force is constant. That is, if we know the net work done on a particle as it undergoes a displacement, we also know the change in its kinetic energy. This is the most important concept in this chapter, so you must understand it thoroughly.

- Recognize that the gravitational potential energy of the mass-Earth system, $PE_g = mgy$, can be positive, negative, or zero, depending on the location of the reference level used to measure y. Be aware of the fact that although PE depends on the origin of the coordinate system, *the change in potential energy,* $(PE)_f - (PE)_i$, *is independent of the coordinate system used to define PE.*

- Calculate average power when work is accomplished over a time interval and instantaneous power when an applied force acts on an object moving with speed v.

SOLUTIONS TO SELECTED END-OF-CHAPTER PROBLEMS

7. A sledge loaded with bricks has a total mass of 18.0 kg and is pulled at constant speed by a rope inclined at 20.0° above the horizontal. The sledge moves a distance of 20.0 m on a horizontal surface. The coefficient of kinetic friction between the sledge and surface is 0.500. (a) What is the tension in the rope? (b) How much work is done by the rope on the sledge? (c) What is the mechanical energy lost due to friction?

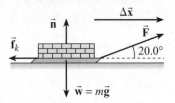

Solution

(a) The sledge has constant velocity, so

$$\Sigma F_y = n + F\sin 20.0° - mg = ma_y = 0 \quad \Rightarrow \quad n = mg - F\sin 20.0°$$

Thus, the kinetic friction force is

$$f_k = \mu_k n = \mu_k \left(mg - F\sin 20.0°\right)$$

Also,

$$\Sigma F_x = F\cos 20.0° - f_k = ma_x = 0$$

or

$$F\cos 20.0° - \mu_k \left(mg - F\sin 20.0°\right) = 0$$

Simplifying this result gives the tension in the rope as

$$F = \frac{\mu_k mg}{\cos 20.0° + \mu_k \sin 20.0°} = \frac{0.500(18.0 \text{ kg})(9.80 \text{ m/s}^2)}{\cos 20.0° + 0.500\sin 20.0°} = 79.4 \text{ N} \qquad \Diamond$$

(b) The work done by the rope on the sledge is $W_F = (F\cos 20.0°)\Delta x$, or

$$W_F = (79.4 \text{ N})(20.0 \text{ m})\cos 20.0° = 1.49 \times 10^3 \text{ J} = 1.49 \text{ kJ}$$ ◊

(c) The friction force and the displacement are in opposite directions ($\theta = 180°$), so the work done by friction on the sledge is

$$W_{f_k} = (f_k \cos 180°)\Delta x = \mu_k (mg - F\sin 20.0°)(\Delta x)\cos 180°$$
$$= 0.500\left[(18.0 \text{ kg})(9.80 \text{ m/s}^2) - (79.4 \text{ N})\sin 20.0°\right](20.0 \text{ m})(-1)$$

or

$$W_{f_k} = -1.49 \times 10^3 \text{ J} = -1.49 \text{ kJ}$$ ◊

13. A 70-kg base runner begins his slide into second base when he is moving at a speed of 4.0 m/s. The coefficient of friction between his clothes and Earth is 0.70. He slides so that his speed is zero just as he reaches the base. (a) How much mechanical energy is lost due to friction acting on the runner? (b) How far does he slide?

Solution

(a) Since the base runner slides on a level surface, the change in gravitational potential energy is

$$\Delta PE_g = mgy_f - mgy_i = mg(\Delta y) = 0$$

The change in mechanical energy is therefore

$$W_{nc} = (KE + PE)_f - (KE + PE)_i = (\Delta KE) + (\Delta PE) = \left(\tfrac{1}{2}mv_f^2 - \tfrac{1}{2}mv_i^2\right) + (0)$$

or

$$W_{nc} = \tfrac{1}{2}m\left(v_f^2 - v_i^2\right) = \frac{1}{2}(70 \text{ kg})\left[0 - (4.0 \text{ m/s})^2\right] = -5.6 \times 10^2 \text{ J}$$ ◊

(b) The friction force is directed opposite to the displacement $(\theta = 180°)$ and has magnitude $f_k = \mu_k n = \mu_k mg$. All other forces acting on the base runner are perpendicular to the displacement and, hence, do no work. The total work done by nonconservative forces is then $W_{nc} = (f_k \cos 180°)\Delta x$, so the displacement is

$$\Delta x = \frac{W_{nc}}{f_k \cos 180°} = \frac{W_{nc}}{-\mu_k mg} = \frac{-5.6 \times 10^2 \text{ J}}{-(0.70)(70 \text{ kg})(9.80 \text{ m/s}^2)} = 1.2 \text{ m}$$ ◊

23. A 2 300-kg pile driver is used to drive a steel beam into the ground. The pile driver falls 7.50 m before coming into contact with the top of the beam, and it drives the beam 18.0 cm farther into the ground as it comes to rest. Using energy considerations, calculate the average force the beam exerts on the pile driver while the pile driver is brought to rest.

Solution

The pile driver is dropped from rest at a height of 7.50 meters *above* the top of the beam and comes to rest again at a distance $d = 0.180$ m *below* the original level of the top of the beam. During this motion, the only nonconservative force acting on the pile driver is an upward force of magnitude F exerted on it by the top of the beam for the very brief duration of the collision. The work done on the pile driver by nonconservative forces during its motion is $W_{nc} = (F \cos 180°) d$. The factor $\cos 180°$ is included because the force $\vec{F}$ and the displacement $\vec{d}$ are in opposite directions.

From the work-energy theorem, this work can also be expressed as

$$W_{nc} = \left(KE_f - KE_i\right) + \left(PE_f - PE_i\right) = \frac{1}{2}m\left(v_f^2 - v_i^2\right) + mg\left(y_f - y_i\right)$$

so

$$W_{nc} = (F \cos 180°)\, d = \frac{1}{2}m\left(v_f^2 - v_i^2\right) + mg\left(y_f - y_i\right)$$

Choosing $y = 0$ at the original level of the top of the beam, the pile driver starts from rest $\left(v_i = 0\right)$ at $y_i = +7.50$ m and comes to rest again $\left(v_f = 0\right)$ at $y_f = -0.180$ m. Therefore,

$$-F(0.180 \text{ m}) = \frac{1}{2}m(0 - 0) + (2\,300 \text{ kg})\left(9.80 \text{ m/s}^2\right)(-0.180 \text{ m} - 7.50 \text{ m})$$

yielding

$$F = \frac{-1.73 \times 10^5 \text{ J}}{-0.180 \text{ m}} = 9.62 \times 10^5 \text{ N directed upward}$$

◊

29. A 50.0-kg projectile is fired at an angle of 30.0° above the horizontal with an initial speed of 1.20×10^2 m/s from the top of a cliff 142 m above level ground, where the ground is taken to be $y = 0$. (a) What is the initial total mechanical energy of the projectile? (b) Suppose the projectile is traveling 85.0 m/s at its maximum height of $y = 427$ m. How much work has been done on the projectile by air friction? (c) What is the speed of the projectile immediately before it hits the ground if air friction does one and a half times as much work on the projectile when it is going down as it did when it was going up?

Solution

(a) Taking $y = 0$, and hence $PE_g = 0$, at ground level, the mechanical energy of the projectile as it leaves the launch point is

$$E_i = KE_i + \left(PE_g\right)_i = \frac{1}{2}mv_i^2 + mgy_i$$

$$= \frac{1}{2}(50.0 \text{ kg})\left(1.20 \times 10^2 \text{ m/s}\right)^2 + (50.0 \text{ kg})\left(9.80 \text{ m/s}^2\right)(142 \text{ m}) = 4.30 \times 10^5 \text{ J} \quad \lozenge$$

(b) If the projectile has a speed of $v_{peak} = 85.0$ m/s when it reaches its maximum altitude at $y_{peak} = 427$ m above ground level, the work done on the projectile as it was rising has been

$$\left(W_{nc}\right)_{rise} = \Delta KE + \Delta PE_g = \frac{1}{2}m\left(v_{peak}^2 - v_i^2\right) + mg\left(y_{peak} - y_i\right)$$

or

$$\left(W_{nc}\right)_{rise} = \frac{1}{2}(50.0 \text{ kg})\left[(85.0 \text{ m/s})^2 - \left(1.20 \times 10^2\right)^2\right]$$

$$+ (50.0 \text{ kg})\left(9.80 \text{ m/s}^2\right)(427 \text{ m} - 142 \text{ m})$$

which yields

$$\left(W_{nc}\right)_{rise} = -3.97 \times 10^4 \text{ J} \quad \lozenge$$

(c) If air friction does one and a half times as much work on the projectile while the projectile falls to the ground as it did while the projectile was rising, the total work done on the projectile by air friction during its flight is

$$\left(W_{nc}\right)_{total} = \left(W_{nc}\right)_{rise} + \left(W_{nc}\right)_{fall} = \left(W_{nc}\right)_{rise} + 1.50\left(W_{nc}\right)_{rise}$$

$$= 2.50\left(W_{nc}\right)_{rise} = 2.50\left(-3.97 \times 10^4 \text{ J}\right) = -9.93 \times 10^4 \text{ J}$$

Applying the work-energy theorem, $(W_{nc})_{total} = (KE_f - KE_i) + (PE_g)_f - (PE_g)_i$, to the full flight of the projectile, we obtain $KE_f = KE_i + (PE_g)_i - (PE_g)_f + (W_{nc})_{total}$

or

$$\frac{1}{2}mv_f^2 = \frac{1}{2}mv_i^2 + mg\left(y_i - y_f\right) + \left(W_{nc}\right)_{total}$$

The speed of the projectile just before it hits the ground is then

$$v_f = \sqrt{v_i^2 + 2g(y_i - y_f) + \frac{2(W_{nc})_{total}}{m}}$$

$$= \sqrt{(1.20 \times 10^2 \text{ m/s})^2 + 2(9.80 \text{ m/s}^2)(142 \text{ m} - 0) + \frac{2(-9.93 \times 10^4 \text{ J})}{50.0 \text{ kg}}}$$

or

$$v_f = 115 \text{ m/s} \qquad \qquad \diamond$$

37. Tarzan swings on a 30.0-m-long vine initially inclined at an angle of 37.0° with the vertical. What is his speed at the bottom of the swing (a) if he starts from rest? (b) if he pushes off with a speed of 4.00 m/s?

Solution

If we ignore air resistance, the only nonconservative force acting on Tarzan, as he swings from the initial position shown to the bottom of his arc, is the tension in the vine. However, this tension is directed toward the center of his circular path, and at each point on his path, is perpendicular to his motion. This means that the tension force does no work on Tarzan, or $W_{nc} = 0$.

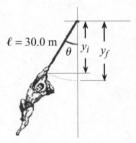

The work-energy theorem then takes the form

$$W_{nc} = (KE_f - KE_i) + (PE_f - PE_i) = 0 \quad \text{or} \quad KE_f = KE_i + (PE_i - PE_f)$$

Since $KE = mv^2/2$, and the potential energy is gravitational, $PE = mgy$, this becomes

$$\frac{1}{2} m v_f^2 = \frac{1}{2} m v_i^2 + mg(y_i - y_f)$$

Realizing that Tarzan's y coordinate is $y = -\ell\cos\theta = -(30.0 \text{ m})\cos\theta$, we see that

$$y_i - y_f = -(30.0 \text{ m})(\cos\theta_i - \cos\theta_f) = -(30.0 \text{ m})(\cos 37.0° - \cos 0°) = +6.04 \text{ m}$$

and his speed at the lowest point on the swing is

$$v_f = \sqrt{v_i^2 + 2g(6.04 \text{ m})}$$

(a) If he starts from rest $(v_i = 0)$,

$$v_f = \sqrt{0 + 2(9.80 \text{ m/s}^2)(6.04 \text{ m})} = 10.9 \text{ m/s} \qquad \diamond$$

(b) If $v_i = 4.00$ m/s,

$$v_f = \sqrt{(4.00 \text{ m/s})^2 + 2(9.80 \text{ m/s}^2)(6.04 \text{ m})} = 11.6 \text{ m/s}$$

◊

47. A skier starts from rest at the top of a hill that is inclined 10.5° with respect to the horizontal. The hillside is 200 m long, and the coefficient of friction between snow and skis is 0.075 0. At the bottom of the hill, the snow is level and the coefficient of friction is unchanged. How far does the skier move along the horizontal portion of the snow before coming to rest?

Solution

Select the reference level of gravitational potential energy ($PE_g = 0$) at the level of the base of the hill and let x represent the horizontal distance the skier travels after reaching this level.

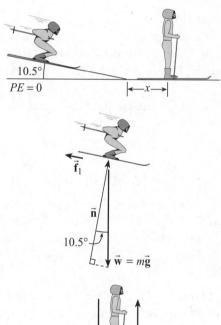

While on the hill, the normal force exerted on the skier by the snow is $n_1 = mg \cos 10.5°$ and the friction force is

$$f_1 = \mu_k n_1 = \mu_k mg \cos 10.5°$$

On the level snow, the normal force is $n_2 = mg$ and the friction force is $f_2 = \mu_k n_2 = \mu_k mg$

Consider the entire trip, from when the skier starts from rest on the hill until the skier comes to rest on the level snow. Then, $y_i - y_f = (200 \text{ m})\sin 10.5°$ and $v_i = v_f = 0$.

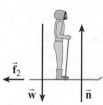

The work done by nonconservative forces is

$$W_{nc} = f_1(200 \text{ m})\cos 180° + (f_2)(x)\cos 180°$$
$$= -\mu_k mg \left[(200 \text{ m})\cos 10.5° + x \right]$$

Application of the work-energy theorem,

$$W_{nc} = \left(KE + PE_g\right)_f - \left(KE + PE_g\right)_i$$

then gives

$$-\mu_k mg \left[(200 \text{ m})\cos 10.5° + x \right] = 0 + mgy_f - 0 - mgy_i = -mg(200 \text{ m})\sin 10.5°$$

or

$$x = (200 \text{ m})\left[\frac{\sin 10.5°}{\mu_k} - \cos 10.5°\right]$$

With $\mu_k = 0.075\ 0$, this yields $x = 289$ m. ◊

---

51. A 3.50-kN piano is lifted by three workers at constant speed to an apartment 25.0 m above the street using a pulley system fastened to the roof of the building. Each worker is able to deliver 165 W of power, and the pulley system is 75.0% efficient (so that 25.0% of the mechanical energy is lost due to friction in the pulley). Neglecting the mass of the pulley, find the time required to lift the piano from the street to the apartment.

Solution

The work that must be done on the piano to lift it at constant speed from the street up to the apartment is given by the work-energy theorem as

$$W_{nc} = \Delta KE + \Delta PE = \frac{1}{2}m\left(v_f^2 - v_i^2\right) + mg\left(y_f - y_i\right) = 0 + mg\left(y_f - y_i\right)$$

or

$$W_{nc} = \left(3.50 \times 10^3 \text{ N}\right)(25.0 \text{ m} - 0) = 8.75 \times 10^4 \text{ J}$$

The three workmen, working together, deliver a total power input to the pulley system of

$$\mathcal{P}_{total} = 3\left(\mathcal{P}_{\substack{single \\ worker}}\right) = 3(165 \text{ W}) = 495 \text{ W} = 495 \text{ J/s}$$

However, the pulley system in use is only 75.0% efficient. This means that 25.0% of the power input is used overcoming friction in the system, and only 75.0% of the power input is transferred to the piano as useful work. The rate at which the system does useful work on the piano is

$$\mathcal{P}_{useful} = 0.750\ \mathcal{P}_{total} = 0.750(495 \text{ J/s}) = 371 \text{ J/s}$$

The time required to lift the piano is then

$$\Delta t = \frac{W_{nc}}{\mathcal{P}_{useful}} = \frac{8.75 \times 10^4 \text{ J}}{371 \text{ J/s}} = 236 \text{ s} = 236 \text{ s}\left(\frac{1 \text{ min}}{60 \text{ s}}\right) = 3.93 \text{ min}$$ ◊

---

59. The force acting on a particle varies as in Figure P5.59. Find the work done by the force as the particle moves (a) from $x = 0$ to $x = 8.00$ m, (b) from $x = 8.00$ m to $x = 10.0$ m, and (c) from $x = 0$ to $x = 10.0$ m.

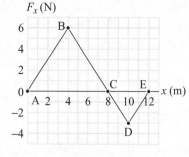

Figure P5.59

Solution

The work done on the particle by the force F as the particle moves from $x = x_i$ to $x = x_f$ is the area under the curve from x_i to x_f.

(a)   For $x = 0$ to $x = 8.00$ m,

$$W = \text{area of triangle } ABC = \frac{1}{2}\overline{AC} \times \text{altitude}$$

$$W_{0 \to 8} = \frac{1}{2}(8.00 \text{ m})(6.00 \text{ N}) = 24.0 \text{ J} \qquad \diamond$$

(b)   For $x = 8.00$ m to $x = 10.0$ m,

$$W_{8 \to 10} = \text{area of triangle } CDE = \frac{1}{2}\overline{CE} \times \text{altitude}$$

$$W_{8 \to 10} = \frac{1}{2}(2.00 \text{ m})(-3.00 \text{ N}) = -3.00 \text{ J} \qquad \diamond$$

(c)   For $x = 0$ to $x = 10.0$ m,

$$W_{0 \to 10} = W_{0 \to 8} + W_{8 \to 10} = 24.0 \text{ J} + (-3.00 \text{ J}) = 21.0 \text{ J} \qquad \diamond$$

69. (a) A 75-kg man steps out a window and falls (from rest) 1.0 m to a sidewalk. What is his speed just before his feet strike the pavement? (b) If the man falls with his knees and ankles locked, the only cushion for his fall is an approximately 0.50-cm give in the pads of his feet. Calculate the average force exerted on him by the ground in this situation. This average force is sufficient to cause cartilage damage in the joints or to break bones.

Solution

(a) While the man is falling, the only force acting on him is the gravitational force (a conservative force). Therefore, the work done by nonconservative forces is zero and the man's mechanical energy (kinetic plus potential) will be conserved. That is:

$$W_{nc} = \left(KE_f - KE_i\right) + \left(PE_f - PE_i\right) = 0 \quad \Rightarrow \quad KE_f + PE_f = KE_i + PE_i$$

Thus,

$$\frac{1}{2}mv_f^2 + mgy_f = \frac{1}{2}mv_i^2 + mgy_i$$

or

$$v_f^2 = v_i^2 + 2g\left(y_i - y_f\right)$$

and, when the man starts from rest, his speed just before touching ground (a fall of 1.0 m) will be

$$v_f = \sqrt{v_i^2 + 2g\left(y_i - y_f\right)} = \sqrt{0 + 2\left(9.8 \text{ m/s}^2\right)(1.0 \text{ m})} = 4.4 \text{ m/s} \qquad \Diamond$$

(b) The instant the pads of the man's feet touch ground, the ground starts exerting an upward nonconservative force, F, on him. Consider the period from this instant until the man comes to rest, after undergoing an additional displacement of $s = 0.50$ cm $= 5.0 \times 10^{-3}$ m (directed at 180° from the direction of F). The work done by nonconservative forces is

$$W_{nc} = \left(F\cos 180°\right)s = \left(KE_f - KE_i\right) + \left(PE_f - PE_i\right)$$

or

$$-Fs = \left(0 - \frac{1}{2}mv_i^2\right) + mg\left(y_f - y_i\right)$$

For this time interval, $v_i = 4.4$ m/s, the final velocity for the interval in part (a), and $(y_f - y_i) = -s$. Thus, the magnitude of the average force is

$$F = m\left(\frac{v_i^2}{2s} + g\right) = (75 \text{ kg})\left(\frac{(4.4 \text{ m/s})^2}{2(5.0 \times 10^{-3} \text{ m})} + 9.8 \text{ m/s}^2\right) = 1.5 \times 10^5 \text{ N} \qquad \Diamond$$

73. A 2.00×10^2-g particle is released from rest at point A on the inside of a smooth hemispherical bowl of radius $R = 30.0$ cm (Fig. P5.73). Calculate (a) its gravitational potential energy at A relative to B, (b) its kinetic energy at B, (c) its speed at B, and (d) its potential energy at C relative to B, and (e) its kinetic energy at C.

Solution

The smooth bowl exerts no friction force on the particle. Therefore, the only nonconservative force acting on the particle as it slides on the inner surface of the hemispherical bowl is a normal force that is always perpendicular to the motion of the particle (and hence, does no work). This means that the total mechanical energy of the particle (kinetic plus potential energies) remains constant:

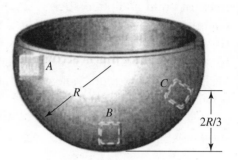

$$W_{nc} = \left(KE_f - KE_i\right) + \left(PE_f - PE_i\right) = 0 \quad \Rightarrow \quad KE_f + PE_f = KE_i + PE_i$$

If we choose $y = 0$ (and hence, $PE_g = mgy = 0$) at the bottom of the bowl (point B), then

$y_A = R = 30.0$ cm and $y_C = 2R/3 = 20.0$ cm $= 0.200$ m.

(a) $\left(PE_g\right)_A = mgy_A = \left(2.00 \times 10^2 \text{ g}\right)\left(\dfrac{1 \text{ kg}}{10^3 \text{ g}}\right)\left(9.80 \dfrac{\text{m}}{\text{s}^2}\right)\left(30.0 \text{ cm}\right)\left(\dfrac{1 \text{ m}}{10^2 \text{ cm}}\right) = 0.588$ J ◊

(b) In going from A to B, with $v_A = 0$ and $y_B = 0$, we have

$$KE_B + \left(PE_g\right)_B = KE_A + \left(PE_g\right)_A$$

or

$$KE_B + 0 = 0 + \left(PE_g\right)_A = 0.588 \text{ J}$$ ◊

(c) $KE_B = \frac{1}{2} m v_B^2 = 0.588$ J, so

$$v_B = \sqrt{\frac{2(0.588 \text{ J})}{m}} = \sqrt{\frac{2(0.588 \text{ J})}{2.00 \times 10^{-1} \text{ kg}}} = 2.42 \text{ m/s}$$ ◊

(d) $\left(PE_g\right)_C = mgy_C = \left(2.00 \times 10^{-1} \text{ kg}\right)\left(9.80 \text{ m/s}^2\right)(0.200 \text{ m}) = 0.392$ J ◊

(e) As the particle goes from B to C, conservation of energy gives

$$KE_C + \left(PE_g\right)_C = KE_B + \left(PE_g\right)_B$$

or

$$KE_C = KE_B + 0 - \left(PE_g\right)_C$$

Thus,

$$KE_C = 0.588 \text{ J} - 0.392 \text{ J} = 0.196 \text{ J}$$ ◊

81. A child's pogo stick (Figure P5.81) stores energy in a spring ($k = 2.50 \times 10^4$ N/m). At position A ($x_1 = -0.100$ m), the spring compression is a maximum and the child is momentarily at rest. At position B ($x = 0$), the spring is relaxed and the child is moving upwards. At position C, the child is again momentarily at rest at the top of the jump. Assuming that the combined mass of child and pogo stick is 25.0 kg, (a) calculate the total energy of the system if both potential energies are zero at $x = 0$, (b) determine x_2, (c) calculate the speed of the child at $x = 0$, (d) determine the value of x for which the kinetic energy of the system is a maximum, and (e) obtain the child's maximum upward speed.

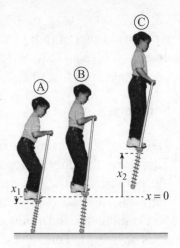

Figure P5.81

Solution

We choose $PE_g = 0$ at the level where the spring is relaxed ($x = 0$), or at the level of position B.

(a) At position A, $v = 0$ and the total energy of the system is given by

$$E = \left(KE + PE_g + PE_s\right)_A = \frac{1}{2}mv_A^2 + mg\,x_1 + \frac{1}{2}k\,x_1^2$$

or

$$E = 0 + (25.0 \text{ kg})(9.80 \text{ m/s}^2)(-0.100 \text{ m}) + \frac{1}{2}(2.50 \times 10^4 \text{ N/m})(-0.100 \text{ m})^2$$

$$E = 101 \text{ J} \qquad \Diamond$$

(b) In position C, $v = 0$ and the spring is uncompressed giving $KE = 0$ and $PE_s = 0$ Hence,

$$E = \left(0 + PE_g + 0\right)_C = mg\,x_2$$

or

$$x_2 = \frac{E}{mg} = \frac{101 \text{ J}}{(25.0 \text{ kg})(9.80 \text{ m/s}^2)} = 0.410 \text{ m} \qquad \Diamond$$

(c) At Position B, $PE_g = PE_s = 0$ and $E = \left(KE + 0 + 0\right)_B = \frac{1}{2}mv_B^2$ Therefore,

$$v_B = \sqrt{\frac{2E}{m}} = \sqrt{\frac{2(101 \text{ J})}{25.0 \text{ kg}}} = 2.84 \text{ m/s} \qquad \Diamond$$

(d) Where the velocity (and hence kinetic energy) is a maximum, the slope of the velocity versus time graph (that is, the acceleration) must be zero. Thus, $\Sigma F_y = ma_y = 0$ or the upward force due to the spring must exactly counterbalance the downward gravitational force where the kinetic energy is a maximum. Taking upward as positive, the position of interest is defined by $\Sigma F_y = -kx - mg = 0$, or

$$x = -\frac{mg}{k} = -\frac{(25.0 \text{ kg})(9.80 \text{ m/s}^2)}{2.50 \times 10^4 \text{ N/m}} = -9.80 \times 10^{-3} \text{ m} = -9.80 \text{ mm} \qquad \Diamond$$

(e) From the total energy, $E = KE + PE_g + PE_s = \frac{1}{2}mv^2 + mgx + \frac{1}{2}kx^2$, we find

$$v = \sqrt{\frac{2E}{m} - 2gx - \frac{k}{m}x^2}$$

Where the speed, and hence kinetic energy, is a maximum (that is, at $x = -9.80 \times 10^{-3}$ m), this gives

$$v_{max} = \left[\frac{2(101 \text{ J})}{(25.0 \text{ kg})} - 2\left(9.80 \ \frac{m}{s^2}\right)(-9.80 \times 10^{-3} \text{ m}) \right.$$
$$\left. -\left(\frac{2.50 \times 10^4 \text{ N/m}}{25.0 \text{ kg}}\right)(-9.80 \times 10^{-3} \text{ m})^2 \right]^{\frac{1}{2}}$$

or

$$v_{max} = 2.85 \text{ m/s} \qquad \Diamond$$

---

87. A loaded ore car has a mass of 950 kg and rolls on rails with negligible friction. It starts from rest and is pulled up a mine shaft by a cable connected to a winch. The shaft is inclined at 30.0° above the horizontal. The car accelerates uniformly to a speed of 2.20 m/s in 12.0 s and then continues at constant speed. (a) What power must the winch motor provide when the car is moving at constant speed? (b) What maximum power must the motor provide? (c) What total energy transfers out of the motor by work by the time the car moves off the end of the track, which is of length 1 250 m?

Solution

(a) When the car is being towed up the incline at constant speed, the tension in the cable is $F = mg\sin\theta$, and the power input from the motor is

$$\mathcal{P} = Fv = mgv\sin\theta$$

or

$$\mathcal{P} = (950 \text{ kg})(9.80 \text{ m/s}^2)(2.20 \text{ m/s})\sin 30.0° = 1.02 \times 10^4 \text{ W} = 10.2 \text{ kW} \qquad \Diamond$$

(b) While the car is accelerating, the tension in the cable is

$$F_a = mg\sin\theta + ma = m\left(g\sin\theta + \frac{\Delta v}{\Delta t}\right)$$

$$= (950 \text{ kg})\left[(9.80 \text{ m/s}^2)\sin 30.0° + \frac{2.20 \text{ m/s} - 0}{12.0 \text{ s}}\right] = 4.83\times10^3 \text{ N}$$

The power input, $\mathcal{P} = Fv$, is a maximum at the instant when both the tension in the cable and the speed of the car have their maximum values. The maximum tension $(F = F_a)$ occurs as the car is being accelerated, and maximum speed $(v_{max} = 2.20 \text{ m/s})$ also occurs at the last instant of the acceleration phase. Thus, the maximum power from the motor is

$$\mathcal{P}_{max} = F_a v_{max} = (4.83\times10^3 \text{ N})(2.20 \text{ m/s}) = 10.6 \text{ kW} \qquad \Diamond$$

(c) The work the motor does moving the car up the frictionless track is

$$W_{nc} = (KE + PE)_f - (KE + PE)_i = \tfrac{1}{2}mv_f^2 + mg(y_f - y_i) = m\left[\tfrac{1}{2}v_f^2 + g(L\sin\theta)\right]$$

or

$$W_{nc} = (950 \text{ kg})\left[\tfrac{1}{2}(2.20 \text{ m/s})^2 + (9.80 \text{ m/s}^2)(1\,250 \text{ m})\sin 30.0°\right] = 5.82\times10^6 \text{ J} \qquad \Diamond$$

6

Momentum and Collisions

NOTES FROM SELECTED CHAPTER SECTIONS

6.1 Momentum and Impulse

The **time rate of change of the momentum** of a particle is equal to the resultant force on the particle. The **impulse** of a force is a vector quantity and is equal to the change in momentum of the particle on which the force acts. The impulse or change in momentum of an object is equal to the area under the force vs. time graph from the beginning to the end of the time interval during which the force is in contact with the object. Under the **impulse approximation,** it is assumed that one of the forces acting on a particle is of short time duration but of much greater magnitude than any of the other forces.

6.2 Conservation of Momentum

6.3 Collisions

The principle of **conservation of linear momentum** can be stated for an isolated system of objects. This is a system on which no external forces (e.g., friction or gravity) are acting. **When no external forces act on a system the total linear momentum of the system remains constant.** Remember, momentum is a vector quantity and the momentum of each individual particle may change. The total momentum of the entire system of particles however will remain constant. A collision between two or more masses is an important example of conservation of momentum.

For **any type of collision,** the total momentum before the collision equals the total momentum just after the collision.

In an **inelastic collision,** the total momentum is conserved; however, the total kinetic energy is not conserved.

In a **perfectly inelastic** collision, the two colliding objects stick together following the collision. This corresponds to a maximum loss in kinetic energy.

In an **elastic collision,** both momentum and kinetic energy are conserved.

6.4 Glancing Collisions

The law of conservation of momentum is not restricted to one-dimensional collisions. If two masses undergo a **two-dimensional** (glancing) **collision** and there are no external forces acting, the total momentum is conserved in both the x- and y-directions independently.

EQUATIONS AND CONCEPTS

Linear momentum of a particle is defined as the product of its mass m and velocity $\vec{v}$. The SI units of linear momentum are kg·m/s. Momentum is a vector quantity; the defining equation can be written in component form.

$$\vec{p} \equiv m\vec{v} \qquad (6.1)$$
$$p_x = mv_x$$
$$p_y = mv_y$$
$$p_z = mv_z$$

The **impulse** imparted to an object is the product of the net force acting on the object and the time interval over which the force acts. *Impulse is a vector quantity and has the same direction as the applied force. The SI unit of impulse is kilogram meter per second.*

$$\vec{I} \equiv \vec{F}\Delta t \qquad (6.4)$$

The **impulse-momentum theorem** states that the impulse acting on an object equals the change in momentum of the object. *Impulse is equal to the area under the force-time graph.*

$$\vec{F}_{av}\,\Delta t = \Delta\vec{p} = m\vec{v}_f - m\vec{v}_i \qquad (6.5)$$

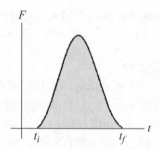

The **time-averaged force** is defined as that constant force which would impart the same impulse to a particle as the actual time-varying force acting over the same time interval. *In the figure, the impulse is equal to the area under the dashed line.*

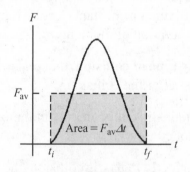

The **principle of conservation of momentum** is stated in Equation (6.7). *When two objects interact in a collision (and exert forces on each other) and no external forces act on the two-object system, the total momentum of the system before the collision equals the total momentum after the collision.*

$$m_1\vec{v}_{1i} + m_2\vec{v}_{2i} = m_1\vec{v}_{1f} + m_2\vec{v}_{2f} \qquad (6.7)$$

Conservation of momentum for a two-body collision. Equation (6.7) can be extended to include any number of objects.

In a **perfectly inelastic collision,** momentum is conserved but kinetic energy is not. *The colliding objects stick together so that they have a common final velocity.*

$$m_1v_{1i} + m_2v_{2i} = (m_1 + m_2)v_f \qquad (6.8)$$

In an **elastic collision,** both momentum and kinetic energy are conserved. *Equations (6.10) and (6.11) apply in the case of "head-on" collisions.*

$$m_1v_{1i} + m_2v_{2i} = m_1v_{1f} + m_2v_{2f} \qquad (6.10)$$

$$\tfrac{1}{2}mv_{1i}^2 + \tfrac{1}{2}mv_{2i}^2 = \tfrac{1}{2}m_{1f}^2 + \tfrac{1}{2}m_2v_{2f}^2 \qquad (6.11)$$

The **relative velocity** before a perfectly elastic collision between two bodies equals the negative of the relative velocity of the two bodies following the collision. Equation (6.14) results from combining Equations (6.10) and (6.11) to eliminate m_1 and m_2. *Equation (6.14) is valid only in elastic collisions.*

$$v_{1i} - v_{2i} = -(v_{1f} - v_{2f}) \qquad (6.14)$$

A **two-dimensional elastic collision** in which an object m_1 moves along the x-axis and collides with m_2 initially at rest is illustrated in the figure below. *Momentum is conserved along each direction.* Angles in Equations (6.15) and (6.16) are defined in the diagram.

x-component: $\qquad\qquad\qquad$ (6.15)

$$m_1v_{1i} + 0 = m_1v_{1f}\cos\theta + m_2v_{2f}\cos\phi$$

y-component: $\qquad\qquad\qquad$ (6.16)

$$0 + 0 = m_1v_{1f}\sin\theta - m_2v_{2f}\sin\phi$$

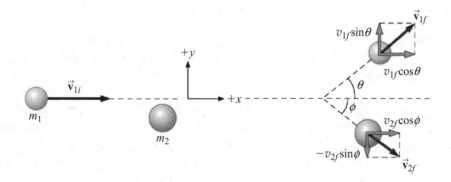

The **equation for rocket propulsion** states that the *change in the speed* of the rocket (as the mass decreases from M_i to M_f) is proportional to the exhaust speed of the ejected gases.

$$v_f - v_i = v_e \ln\left(\frac{M_i}{M_f}\right) \qquad (6.19)$$

The **thrust on a rocket** is the force exerted on the rocket by the ejected exhaust gases. The thrust increases as both the exhaust velocity and the burn rate increase,

$$\text{Thrust} = \left| v_e \frac{\Delta M}{\Delta t} \right| \qquad (6.20)$$

SUGGESTIONS, SKILLS, AND STRATEGIES

The following procedure is recommended when dealing with problems involving collisions between two objects:

1. Set up a coordinate system and define velocities with respect to that system. That is, objects moving in the direction selected as the positive direction of the x-axis are considered as having a positive velocity and negative if moving in the negative x-direction. It is convenient to have the x-axis coincide with the direction of one of the initial velocities.

2. Show all velocity vectors with labels and include all the given information including scattering angles.

3. Write expressions for the momentum of each object before and after the collision. (In two-dimensional collision problems, write expressions for the x- and y-components of momentum before and after the collision.) Remember to include the appropriate signs for their velocity directions.

4. Now write expressions for the total momentum before and the total momentum after the collision and equate the two. (For two-dimensional collisions, separate expressions must be written for the momentum in the x- and y-directions. See Equations (6.15) and (6.16). *It is important to emphasize that it is the momentum of the system that is conserved, not the momentum of the individual objects.*

5. If the collision is perfectly inelastic (kinetic energy is not conserved and the two objects have a common velocity following the collision) you should then proceed to solve the momentum equations for the unknown quantities.

6. If the collision is elastic, kinetic energy is also conserved, so you can equate the total kinetic energy before the collision to the total kinetic energy after the collision. This gives an additional relationship between the various velocities. The conservation of kinetic energy for elastic collisions leads to the expression $v_{1i} - v_{2i} = -(v_{1f} - v_{2f})$, which is often easier to use in solving elastic collision problems than is an expression for conservation of kinetic energy.

REVIEW CHECKLIST

- The impulse of a force acting on a particle during some time interval equals the change in momentum of the particle, and the impulse equals the area under the force vs. time graph.

- The momentum of any isolated system (or any system for which the net external force is zero) is conserved, regardless of the nature of the forces between the masses that comprise the system.

- There are two types of collisions that can occur between particles, namely elastic and inelastic collisions. Recognize that a perfectly inelastic collision is an inelastic collision in which the colliding particles stick together after the collision, and hence move as a composite particle. Kinetic energy is conserved in elastic collisions, but not in the case of inelastic collisions. Linear momentum is conserved in both elastic and inelastic collisions, if the net external force acting on the system of colliding objects is zero.

- The conservation of linear momentum applies not only to head-on collisions (one-dimensional), but also to glancing collisions (in two or three dimensions). For example, in a two-dimensional collision, the total momentum component in the x-direction is conserved and the total momentum component in the y-direction is conserved.

- The equations for conservation of momentum and kinetic energy can be used to calculate the final velocities in a two-body head-on elastic collision. For a perfectly inelastic collision, the equation of conservation of momentum may be used to calculate the final velocity of the composite particle.

SOLUTIONS TO SELECTED END-OF-CHAPTER PROBLEMS

3. A pitcher claims he can throw a 0.145-kg baseball with as much momentum as a 3.00-g bullet moving with a speed of 1.50×10^3 m/s. (a) What must the baseball's speed be if the pitcher's claim is valid? (b) Which has greater kinetic energy, the ball or the bullet?

Solution

(a) If the pitcher's claim is correct, then $p_{ball} = p_{bullet}$, or

$$m_{ball} v_{ball} = m_{bullet} v_{bullet}$$

and the speed of the baseball must be

$$v_{ball} = \frac{m_{bullet} v_{bullet}}{m_{ball}} = \frac{\left(3.00 \times 10^{-3} \text{ kg}\right)\left(1.50 \times 10^3 \text{ m/s}\right)}{0.145 \text{ kg}} = 31.0 \text{ m/s} \qquad \lozenge$$

(b)   The kinetic energy of the bullet is

$$KE_{bullet} = \frac{1}{2}m_{bullet}v_{bullet}^2 = \frac{(3.00\times10^{-3}\text{ kg})(1.50\times10^3\text{ m/s})^2}{2} = 3.38\times10^3\text{ J}$$

and the kinetic energy of the baseball is

$$KE_{ball} = \frac{1}{2}m_{ball}v_{ball}^2 = \frac{(0.145\text{ kg})(31.0\text{ m/s})^2}{2} = 69.7\text{ J}$$

The bullet has the larger kinetic energy by a factor of

$$\frac{KE_{bullet}}{KE_{ball}} = \frac{3.38\times10^3\text{ J}}{69.7\text{ J}} = 48.5 \qquad \Diamond$$

9.   A 0.280-kg volleyball approaches a player horizontally with a speed of 15.0 m/s. The player strikes the ball with her fist and causes the ball to move in the opposite direction with a speed of 22.0 m/s. (a) What impulse is delivered to the ball by the player? (b) If the player's fist is in contact with the ball for 0.060 0 s, find the magnitude of the average force exerted on the player's fist.

Solution

The sketch at the right shows the volleyball immediately before and immediately after it is hit by the player's fist. Observe that we have arbitrarily chosen the direction of the ball's final velocity to be the positive horizontal direction.

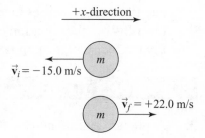

(a)   The impulse delivered to an object over some time interval is just equal to the change in the object's momentum during that interval. That is:

$$\vec{\mathbf{I}} = \Delta\vec{\mathbf{p}} = m\vec{\mathbf{v}}_f - m\vec{\mathbf{v}}_i$$

so

$$\vec{\mathbf{I}} = (0.280\text{ kg})(+22.0\text{ m/s}) - (0.280\text{ kg})(-15.0\text{ m/s})$$

and

$$\vec{\mathbf{I}} = +6.16\text{ kg}\cdot\text{m/s} - (-4.20\text{ kg}\cdot\text{m/s}) = +10.4\text{ kg}\cdot\text{m/s}$$

or

$$\vec{\mathbf{I}} = 10.4\text{ kg}\cdot\text{m/s in the direction of the ball's final velocity} \qquad \Diamond$$

(b) The impulse delivered to an object in a time interval Δt can also be expressed as $\vec{I} = \vec{F}_{av}(\Delta t)$, where $\vec{F}_{av}$ is the average force exerted on the object during that interval. The average force exerted on the ball during the $\Delta t = 0.60\,0$ s time interval that it is in contact with the player's fist is then

$$\vec{F}_{av} = \frac{\vec{I}}{\Delta t} = \frac{+10.4 \ \text{kg} \cdot \text{m/s}}{0.060\,0 \ \text{s}} = +173 \ \text{kg} \cdot \text{m/s}^2 = +173 \ \text{N}$$

Thus, the fist exerts an average force of 173 N in the direction of the ball's final velocity on the ball. By Newton's third law, the ball must have exerted an average force of 173 N in the opposite direction (the direction of the ball's initial velocity) on the player's fist. ◊

---

19. The front 1.20 m of a 1 400-kg car is designed as a "crumple zone" that collapses to absorb the shock of a collision. If a car traveling 25.0 m/s stops uniformly in 1.20 m, (a) how long does the collision last, (b) what is the magnitude of the average force on the car, and (c) what is the acceleration of the car? Express the acceleration as a multiple of the acceleration of gravity.

Solution

(a) The duration of the collision is the time required for the car to travel the 1.20 m length of the "crumple zone." Assuming a uniform deceleration of the car, this time is given by

$$\Delta t = \frac{\Delta x}{v_{av}} = \frac{\Delta x}{\left(v_f + v_i\right)/2} = \frac{2(1.20 \ \text{m})}{0 + 25.0 \ \text{m/s}} = 9.60 \times 10^{-2} \ \text{s} \qquad ◊$$

(b) From the Impulse-momentum theorem, $\vec{I} = \vec{F}_{av}\Delta t = \Delta \vec{p}$, the magnitude of the average force exerted on the car during the impact is

$$F_{av} = \frac{|\Delta \vec{p}|}{\Delta t} = \frac{m|\vec{v}_f - \vec{v}_i|}{\Delta t} = \frac{(1\,400 \ \text{kg})|0 - 25.0 \ \text{m/s}|}{9.60 \times 10^{-2} \ \text{s}} = 3.65 \times 10^5 \ \text{N} \qquad ◊$$

(c) The magnitude of the average acceleration of the car during this collision is

$$a_{av} = \frac{|\vec{v}_f - \vec{v}_i|}{\Delta t} = \frac{|0 - 25.0 \ \text{m/s}|}{9.60 \times 10^{-2} \ \text{s}} = 260 \ \text{m/s}^2 = \left(260 \ \text{m/s}^2\right)\left(\frac{1 \ g}{9.80 \ \text{m/s}^2}\right) = 26.6 \ g \qquad ◊$$

---

23. A 45.0-kg girl is standing on a 150-kg plank. The plank, originally at rest, is free to slide on a frozen lake, which is a flat, frictionless surface. The girl begins to walk along the plank at a constant velocity of 1.50 m/s to the right relative to the plank. (a) What is her velocity relative to the surface of the ice? (b) What is the velocity of the plank relative to the surface of the ice?

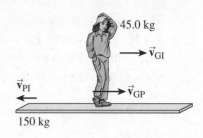

Solution

(a) The velocity of the girl relative to the ice, $\vec{\mathbf{v}}_{GI}$, is $\vec{\mathbf{v}}_{GI} = \vec{\mathbf{v}}_{GP} + \vec{\mathbf{v}}_{PI}$ where $\vec{\mathbf{v}}_{GP}$ = velocity of the girl relative to the plank and $\vec{\mathbf{v}}_{PI}$ = velocity of the plank relative to the ice.

Since we are given that $\vec{\mathbf{v}}_{GP} = 1.50$ m/s, this becomes $\vec{\mathbf{v}}_{GI} = 1.50$ m/s $+ \vec{\mathbf{v}}_{PI}$.　　　**(1)**

Conservation of momentum gives $m_G \vec{\mathbf{v}}_{GI} + m_P \vec{\mathbf{v}}_{PI} = 0$, or

$$\vec{\mathbf{v}}_{PI} = -\left(\frac{m_G}{m_P}\right)\vec{\mathbf{v}}_{GI}\qquad\qquad\text{(2)}$$

Equation (1) becomes $(1 + m_G/m_P)\vec{\mathbf{v}}_{GI} = +1.50$ m/s, or

$$\vec{\mathbf{v}}_{GI} = \frac{+1.50 \text{ m/s}}{1 + \left(\dfrac{45.0 \text{ kg}}{150 \text{ kg}}\right)} = +1.15 \text{ m/s}\qquad\qquad\Diamond$$

(b) Then, using Equation (2),

$$\vec{\mathbf{v}}_{PI} = -\left(\frac{45.0 \text{ kg}}{150 \text{ kg}}\right)(1.15 \text{ m/s}) = -0.346 \text{ m/s}$$

or $\vec{\mathbf{v}}_{PI} = 0.346$ m/s directed opposite to the girl's motion.　　　$\Diamond$

31. Gayle runs at a speed of 4.00 m/s and dives on a sled, initially at rest on the top of a frictionless, snow-covered hill. After she has descended a vertical distance of 5.00 m, her brother, who is initially at rest, hops on her back, and they continue down the hill together. What is their speed at the bottom of the hill if the total vertical drop is 15.0 m? Gayle's mass is 50.0 kg, the sled has a mass of 5.00 kg, and her brother has a mass of 30.0 kg.

Solution

When Gayle dives onto the sled, conservation of momentum gives

$$m_{Gayle+sled}v_2 = m_{Gayle}v_1 + m_{sled}v_0$$

or

$$v_2 = \frac{m_{Gayle}v_1 + m_{sled}(0)}{m_{Gayle+sled}} = \left(\frac{m_{Gayle}}{m_{Gayle+sled}}\right)v_1$$

so Gayle and the sled start down the hill with an initial speed of

$$v_2 = \left(\frac{50.0 \text{ kg}}{55.0 \text{ kg}}\right)(4.00 \text{ m/s}) = 3.64 \text{ m/s}$$

After Gayle and the sled have undergone a 5.00 m decrease in altitude, conservation of mechanical energy gives $KE_f = KE_i + \left(PE_i - PE_f\right)$, or

$$\frac{1}{2}m_{Gayle+sled}v_3^2 = \frac{1}{2}m_{Gayle+sled}v_2^2 + m_{Gayle+sled}g(y_2 - y_3)$$

and the speed down the incline at this point is

$$v_3 = \sqrt{v_2^2 + 2g(y_2 - y_3)} = \sqrt{(3.64 \text{ m/s})^2 + 2(9.80 \text{ m/s}^2)(5.00 \text{ m})} = \sqrt{111} \text{ m/s}$$

When her brother, initially at rest, drops onto her back, conservation of momentum gives speed of the combined system down the slope immediately after impact.

$$\left(m_{Gayle} + m_{sled} + m_{brother}\right)v_4 = m_{Gayle+sled}v_3 + m_{brother}(0)$$

or

$$v_4 = \left(\frac{m_{Gayle+sled}}{m_{Gayle} + m_{sled} + m_{brother}}\right)v_3 = \left(\frac{55.0 \text{ kg}}{50.0 \text{ kg} + 5.00 \text{ kg} + 30.0 \text{ kg}}\right)\left(\sqrt{111} \, \frac{\text{m}}{\text{s}}\right) = 6.82 \text{ m/s}$$

Finally, after the combined system undergoes an additional 10.0 m drop in altitude, conservation of mechanical energy gives

$$\frac{1}{2}m_{total}v_5^2 = \frac{1}{2}m_{total}v_4^2 + m_{total}g(y_4 - y_5)$$

and the speed at the bottom of the hill is

$$v_5 = \sqrt{v_4^2 + g(y_4 - y_5)} = \sqrt{(6.82 \text{ m/s})^2 + 2(9.80 \text{ m/s}^2)(10.0 \text{ m})} = 15.6 \text{ m/s} \qquad \Diamond$$

37. In a Broadway performance, an 80.0-kg actor swings from a 3.75-m-long cable that is horizontal when he starts. At the bottom of his arc, he picks up his 55.0-kg costar in an inelastic collision. What maximum height do they reach after their upward swing?

Solution

The leftmost part of the sketch shows the situation from when the actor starts from rest until just before his impact with his costar. The rightmost part shows the period from just after impact until they come to rest momentarily at the end of the swing.

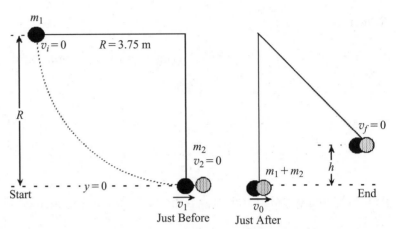

The cable's tension is always perpendicular to the motion, and does no work. Thus, mechanical energy is conserved during the actor's downswing before collision. This gives

$$KE_f = KE_i + \left(PE_i - PE_f\right)$$

or

$$\frac{1}{2} m_1 v_1^2 = \frac{1}{2} m_1 v_0^2 + m_1 g\left(y_i - y_f\right)$$

and his speed just before impact is $v_1 = \sqrt{0 + 2gR} = \sqrt{2\left(9.80 \text{ m/s}^2\right)\left(3.75 \text{ m}\right)} = 8.57 \text{ m/s}$.

Conservation of momentum during the collision with his initially stationary costar gives

$$\left(m_1 + m_2\right)v_a = m_1 v_1 + m_2\left(0\right)$$

and

$$v_a = \frac{m_1 v_1}{m_1 + m_2} = \frac{\left(80.0 \text{ kg}\right)\left(8.57 \text{ m/s}\right)}{80.0 \text{ kg} + 55.0 \text{ kg}} = 5.08 \text{ m/s}$$

is the speed of the pair immediately after the perfectly inelastic collision.

Mechanical energy is again conserved as the actor and costar swing together to height h before coming to rest momentarily. Thus, $KE_f + PE_f = KE_i + PE_i$ or $KE_f + PE_f - PE_i = KE_i$ gives

$$\frac{1}{2}\left(m_1 + m_2\right)\left(0\right) + \left(m_1 + m_2\right)g\left(y_f - y_i\right) = \frac{1}{2}\left(m_1 + m_2\right)v_a^2$$

and the maximum height reached after their upward swing is

$$h = \left(y_f - y_i\right) = \frac{v_a^2}{2g} = \frac{\left(5.08 \text{ m/s}\right)^2}{2\left(9.80 \text{ m/s}^2\right)} = 1.32 \text{ m}$$

◊

40. An 8.00-g bullet is fired into a 250-g block that is initially at rest at the edge of a table of height 1.00 m (Fig. P6.40). The bullet remains in the block, and after the impact the block lands 2.00 m from the bottom of the table. Determine the initial speed of the bullet.

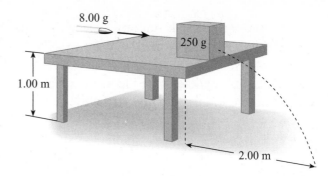

Figure P6.40

Solution

First, consider the motion of the block-bullet combination for the time interval from the instant just after the bullet embeds itself in the block until the block hits the floor. During this interval, the combination is a projectile with an initial velocity that is horizontal ($v_{0y} = 0$) with magnitude v_{0x}. The time the combination is in flight is found from $\Delta y = v_{0y}t + \frac{1}{2}a_y t^2$, which gives

$$-1.00 \text{ m} = 0 + \frac{1}{2}\left(-9.80 \text{ m/s}^2\right)t^2$$

or

$$t = \sqrt{\frac{2(-1.00 \text{ m})}{-9.80 \text{ m/s}^2}} = 0.452 \text{ s}$$

Taking the positive x-direction to be the direction of the bullet's motion in Figure P6.40, the velocity of the block-bullet combination immediately after the bullet collides with the block is

$$V = v_{0x} = \frac{\Delta x}{t} = \frac{2.00 \text{ m}}{0.452 \text{ s}} = 4.43 \text{ m/s}$$

Now apply conservation of momentum from before to just after the bullet-block collision:

$$p_i = p_f \quad \Rightarrow \quad \left(p_{\text{bullet}}\right)_i + \left(p_{\text{block}}\right)_i = \left(p_{\text{combination}}\right)_f$$

or

$$mv_{\text{bullet}} + M(0) = (m + M)V$$

giving

$$v_{\text{bullet}} = \left(1 + \frac{M}{m}\right)V = \left(1 + \frac{250 \text{ g}}{8.00 \text{ g}}\right)(4.43 \text{ m/s}) = 143 \text{ m/s} \qquad \lozenge$$

Observe that

$$v_{\text{bullet}} = (143 \text{ m/s})\left(\frac{1 \text{ mi/h}}{0.447 \text{ m/s}}\right) \approx 320 \text{ mph}$$

47. A 90.0-kg fullback running east with a speed of 5.00 m/s is tackled by a 95.0-kg opponent running north with a speed of 3.00 m/s. (a) Why does the tackle constitute a perfectly inelastic collision? (b) Calculate the velocity of the players immediately after the tackle, and (c) determine the mechanical energy that is lost as a result of the collision. Where did the lost energy go?

Solution

(a) Over the short time interval of the collision, external forces have no time to impart significant impulse to the players. The two players move together as a unit after the tackle, so the collision is completely inelastic. ◊

(b) We choose east to be the positive x-direction as shown in the sketch, and conserve both x- and y-components of momentum:

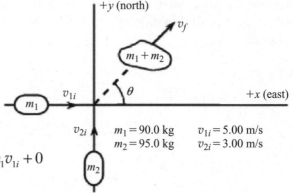

For the eastward x-direction, we have

$$P_{xf} = \Sigma p_{xi} \quad \Rightarrow \quad (m_1 + m_2)v_f \cos\theta = m_1 v_{1i} + 0$$

or

$$v_f \cos\theta = \frac{m_1 v_{1i}}{(m_1 + m_2)} = \frac{(90.0 \text{ kg})(5.00 \text{ m/s})}{90.0 \text{ kg} + 95.0 \text{ kg}}$$

and $v_f \cos\theta = 2.43$ m/s.

For the northward y-direction:

$$P_{yf} = \Sigma p_{yi} \quad \Rightarrow \quad (m_1 + m_2)v_f \sin\theta = 0 + m_2 v_{2i}$$

giving

$$v_f \sin\theta = \frac{m_2 v_{2i}}{(m_1 + m_2)} = \frac{(95.0 \text{ kg})(3.00 \text{ m/s})}{90.0 \text{ kg} + 95.0 \text{ kg}}$$

and

$$v_f \sin\theta = 1.54 \text{ m/s}$$

Therefore,

$$v_f^2 (\sin^2\theta + \cos^2\theta) = v_f^2 = (1.54 \text{ m/s})^2 + (2.43 \text{ m/s})^2$$

and

$$v_f = \sqrt{8.28 \text{ m}^2/\text{s}^2} = 2.88 \text{ m/s}$$

Also,

$$\tan\theta = \frac{v_{fx}\sin\theta}{v_{fx}\cos\theta} = \frac{1.54 \text{ m/s}}{2.43 \text{ m/s}} = 0.633$$

and $\theta = \tan^{-1}(0.633) = 32.3°$.

Thus,

$$\vec{v}_f = 2.88 \text{ m/s at } 32.3° \text{ north of east} \qquad \Diamond$$

(c) $\quad KE_{\text{lost}} = KE_i - KE_f = \dfrac{1}{2}m_1 v_{1i}^2 + \dfrac{1}{2}m_2 v_{2i}^2 - \dfrac{1}{2}(m_1 + m_2)v_f^2$

$$= \frac{1}{2}\left[(90.0 \text{ kg})(5.00 \text{ m/s})^2 + (95.0 \text{ kg})(3.00 \text{ m/s})^2\right] - \frac{1}{2}(185 \text{ kg})(2.88 \text{ m/s})^2$$

or, the kinetic energy lost in the collision is $KE_{\text{lost}} = 785$ J

The lost kinetic energy is transformed into other forms of energy, such as thermal energy and sound.

---

55. A 2.0-g particle moving at 8.0 m/s makes a perfectly elastic head-on collision with a resting 1.0-g object. (a) Find the speed of each particle after the collision. (b) Find the speed of each particle after the collision if the stationary particle has a mass of 10 g. (c) Find the final kinetic energy of the incident 2.0-g particle in the situations described in (a) and (b). In which case does the incident particle lose more kinetic energy?

Solution

Note that the initial velocity of the target object is zero ($v_{2i} = 0$).

From conservation of momentum,

$$m_1 v_{1f} + m_2 v_{2f} = m_1 v_{1i} + 0 \qquad (1)$$

For head-on elastic collisions,

$$v_{1i} - 0 = -\left(v_{1f} - v_{2f}\right) \implies v_{2f} = v_{1f} + v_{1i} \qquad (2)$$

Substituting Equation (2) into Equation (1) and solving for v_{1f} gives

$$v_{1f} = \left(\frac{m_1 - m_2}{m_1 + m_2}\right)v_{1i}$$

Substituting this result into Equation (2) and simplifying yields

$$v_{2f} = \left(\frac{2m_1}{m_1 + m_2}\right)v_{1i}$$

(a) If $m_1 = 2.0$ g, $m_2 = 1.0$ g, and $v_{1i} = 8.0$ m/s, then

$$v_{1f} = \frac{8}{3} \text{ m/s and } v_{2f} = \frac{32}{3} \text{ m/s} \qquad \Diamond$$

(b) If $m_1 = 2.0$ g, $m_2 = 10$ g, and $v_{1i} = 8.0$ m/s, we find

$$v_{1f} = -\frac{16}{3} \text{ m/s and } v_{2f} = \frac{8}{3} \text{ m/s} \qquad \Diamond$$

(c) The final kinetic energy of the 2.0 g particle in each case is

$$\text{Case (a): } KE_{1f} = \frac{1}{2}m_1 v_{1f}^2 = \frac{1}{2}\left(2.0 \times 10^{-3} \text{ kg}\right)\left(\frac{8}{3} \text{ m/s}\right)^2 = 7.1 \times 10^{-3} \text{ J} \qquad \Diamond$$

$$\text{Case (b): } KE_{1f} = \frac{1}{2}m_1 v_{1f}^2 = \frac{1}{2}\left(2.0 \times 10^{-3} \text{ kg}\right)\left(-\frac{16}{3} \text{ m/s}\right)^2 = 2.8 \times 10^{-2} \text{ J} \qquad \Diamond$$

Since the incident kinetic energy is the same in cases (a) and (b), we observe that the incident particle loses more kinetic energy in case (a). $\qquad \Diamond$

63. A 0.500-kg block is released from rest at the top of a frictionless track 2.50 m above the top of a table. It then collides elastically with a 1.00-kg object that is initially at rest on the table, as shown in Figure P6.63. (a) Determine the velocities of the two objects just after the collision. (b) How high up the track does the 0.500-kg object travel back after the collision? (c) How far away from the bottom of the table does the 1.00-kg object land, given that the table is 2.00 m high? (d) How far away from the bottom of the table does the 0.500-kg object eventually land?

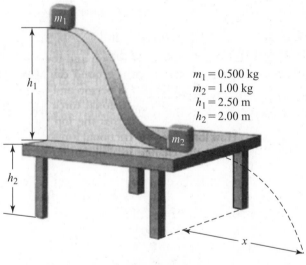

$m_1 = 0.500$ kg
$m_2 = 1.00$ kg
$h_1 = 2.50$ m
$h_2 = 2.00$ m

Figure P6.63

Solution

(a) Since the track is frictionless, we may use conservation of mechanical energy to find the speed of m_1 just before collision.

$$\tfrac{1}{2}m_1 v_1^2 + m_1 g y_{\text{bottom}} = \tfrac{1}{2}m_1 v_0^2 + m_1 g y_{\text{top}} \text{ with } v_0 = 0 \text{ becomes } v_1 = \sqrt{2g\left(y_{\text{top}} - y_{\text{bottom}}\right)} \text{ and}$$

gives $v_1 = \sqrt{2gh_1} = \sqrt{2\left(9.80 \text{ m/s}^2\right)\left(2.50 \text{ m}\right)} = 7.00$ m/s.

From conservation of momentum during the collision and the fact that the velocity of m_2 before impact is $v_2 = 0$, we obtain $m_1 v_{1f} + m_2 v_{2f} = m_1 v_1 + m_2 v_2$, or

$$(0.500 \text{ kg}) v_{1f} + (1.00 \text{ kg}) v_{2f} = (0.500 \text{ kg})(7.00 \text{ m/s}) + 0$$

which simplifies to

$$v_{1f} + 2 v_{2f} = 7.00 \text{ m/s} \qquad \qquad \textbf{(1)}$$

For elastic head-on collisions such as this, $v_1 - v_2 = -(v_{1f} - v_{2f})$. With $v_2 = 0$, and $v_1 = 7.00$ m/s, this yields

$$v_{1f} - v_{2f} = -7.00 \text{ m/s} \qquad \qquad \textbf{(2)}$$

Subtracting Equation (2) from (1) gives $3 v_{2f} = 14.0$ m/s and $v_{2f} = 4.67$ m/s.   ◊

Substituting this result into either (1) or (2) gives $v_{1f} = -2.33$ m/s.   ◊

The negative sign in the last results means that m_1 rebounds back up the ramp.

(b) After the collision, m_1 rebounds up the ramp, starting with kinetic energy $KE_1' = m_1 v_{1f}^2 / 2$ at the bottom. It continues up the frictionless ramp until this initial kinetic energy has been converted into gravitational potential energy, or until $mgh_1' = m_1 v_{1f}^2 / 2$. Thus, the rebound height is

$$h_1' = \frac{v_{1f}^2}{2g} = \frac{(-2.33 \text{ m/s})^2}{2(9.80 \text{ m/s}^2)} = 0.277 \text{ m} \qquad \qquad ◊$$

(c) From $\Delta y = v_{0y} t + \frac{1}{2} a_y t^2$, with $v_{0y} = 0$ and $a_y = -g$, the time required for an object leaving the table horizontally to drop 2.00 m to the floor is

$$t = \sqrt{\frac{2(\Delta y)}{a_y}} = \sqrt{\frac{2(-2.00 \text{ m})}{-9.80 \text{ m/s}^2}} = 0.639 \text{ s}$$

The horizontal distance m_2 travels during its flight to the floor is given by $\Delta x = v_{0x} t$ as

$$x_2 = v_{2f} t = (4.67 \text{ m/s})(0.639 \text{ s}) = 2.98 \text{ m} \qquad \qquad ◊$$

(d) After rebounding up the ramp to a height $h_1' = 0.277$ m, the 0.500-kg object comes to rest momentarily before sliding back down, converting its energy from gravitational potential energy back into kinetic energy. When it arrives at the bottom, all its energy is again in the form of kinetic energy, and it leaves the table with a horizontal speed of $v_{1f}' = |v_{1f}| = 2.33$ m/s. As shown in (c) above, the time for it to drop the 2.00 m to the floor is $t = 0.639$ s, and the horizontal distance it travels from the table as it drops is

$$x_1 = v_{1f}' t = (2.33 \text{ m/s})(0.639 \text{ s}) = 1.49 \text{ m} \qquad \qquad ◊$$

67. A cannon is rigidly attached to a carriage, which can move along horizontal rails, but is connected to a post by a large spring, initially unstretched and with force constant $k = 2.00 \times 10^4$ N/m, as in Figure P6.67. The cannon fires a 200-kg projectile at a velocity of 125 m/s directed 45.0° above the horizontal. (a) If the mass of the cannon and its carriage is 5 000 kg, find the recoil speed of the cannon. (b) Determine the maximum extension of the spring. (c) Find the maximum force the spring exerts on the carriage. (d) Consider the system consisting of the cannon, carriage, and shell. Is the momentum of this system conserved during the firing? Why or why not?

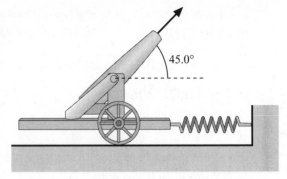

Figure P6.67

Solution

(a) The spring does not have time to stretch during the very brief interval from just before the cannon firing to the instant just after the firing. Thus, without any horizontal, *external* forces acting on the system (cannon + spring + shell) during this time interval, the horizontal component of momentum must be conserved. From this, and taking toward the right as positive, we obtain

$$\left(\Sigma p_x\right)_f = \left(\Sigma p_x\right)_i \quad \text{or} \quad m_{shell}\left(v_{shell}\cos 45.0°\right) + m_{cannon} v_{recoil} = 0 + 0$$

which gives the recoil velocity of the cannon as

$$v_{recoil} = -\left(\frac{m_{shell}}{m_{cannon}}\right) v_{shell}\cos 45.0° = -\left(\frac{200\ \text{kg}}{5\,000\ \text{kg}}\right)(125\ \text{m/s})\cos 45.0° = -3.54\ \text{m/s}$$

and the recoil speed is $\left|v_{recoil}\right| = 3.54$ m/s.   ◊

(b) As the cannon recoils and the spring stretches, no external forces do work on the cannon plus spring system. Therefore, we make use of conservation of mechanical energy to obtain

$$\left(KE + PE_g + PE_s\right)_f = \left(KE + PE_g + PE_s\right)_i \quad \text{or} \quad 0 + 0 + \frac{1}{2}kx_{max}^2 = \frac{1}{2}m_{cannon}\,v_{recoil}^2 + 0 + 0$$

and find

$$x_{max} = \sqrt{\frac{m_{cannon}\,v_{recoil}^2}{k}} = \sqrt{\frac{(5\,000\ \text{kg})(-3.54\ \text{m/s})^2}{2.00 \times 10^4\ \text{N/m}}} = 1.77\ \text{m}$$ ◊

(c) The spring exerts maximum force (directed toward the right) when it has the maximum extension of x_{max}. The magnitude of this maximum force is

$$|F_{max}| = kx_{max} = (2.00 \times 10^4 \text{ N/m})(1.77 \text{ m}) = 3.54 \times 10^4 \text{ N} \qquad \Diamond$$

(d) The vertical component of the system's (cannon + spring + shell) momentum *is not conserved* during the firing of the cannon. The reason for this is that the rail exerts a vertical external force (the normal force) on the cannon and prevents it from recoiling in the vertical direction as the shell starts up the cannon's barrel. However, the spring does not have time to stretch during the cannon firing. Thus, the spring does not yet exert a force on the post, and the post does not yet exert a horizontal reaction force on the system. Without any horizontal external forces present, *the horizontal component of the system's momentum is conserved* during the brief instant from just before to just after the firing. This was the principle we made use of in part (a). $\qquad \Diamond$

---

73. A tennis ball of mass 57.0 g is held just above a basketball of mass 590 g. With their centers vertically aligned, both balls are released from rest at the same time, to fall through a distance of 1.20 m, as shown in Figure P6.73. (a) Find the magnitude of the downward velocity with which the basketball reaches the ground. (b) Assume that an elastic collision with the ground instantaneously reverses the velocity of the basketball while the tennis ball is still moving down. Next, the two balls meet in an elastic collision. To what height does the tennis ball rebound?

Figure P6.73

Solution

(a) When released, both balls start from rest and fall freely under the influence of gravity for 1.20 m before the basketball touches the ground. Their downward speed just before the basketball hits ground may be found from $v_y^2 = v_{0y}^2 + 2a_y\Delta y$. This yields

$$v_i = \sqrt{v_{0y}^2 + 2a_y\Delta y} = \sqrt{0 + 2(-9.80 \text{ m/s}^2)(-1.20 \text{ m})} = 4.85 \text{ m/s} \qquad \Diamond$$

(b) Immediately after the basketball rebounds from the floor, it and the tennis ball meet in an elastic collision. The vertical velocity of each ball just before the collision is

For the tennis ball: $\qquad v_{1i} = -v_i = -4.85 \text{ m/s}$

For the basketball: $\qquad v_{2i} = +v_i = +4.85 \text{ m/s}$

The velocity of the tennis ball immediately after this elastic collision may be found as follows:

Conservation of momentum gives

$$(57.0 \text{ g})v_{1f} + (590 \text{ g})v_{2f} = (57.0 \text{ g})(-4.85 \text{ m/s}) + (590 \text{ g})(+4.85 \text{ m/s})$$

which reduces to

$$v_{2f} = \left(\frac{533}{590}\right)(4.85 \text{ m/s}) - \left(9.66 \times 10^{-2}\right)v_{1f} \tag{1}$$

From the characteristics of a head-on elastic collision, we know that

$$v_{1f} - v_{2f} = -\left(v_{1i} - v_{2i}\right) = -\left(-4.85 \text{ m/s} - 4.85 \text{ m/s}\right) = +9.70 \text{ m/s}$$

or

$$v_{1f} = 9.70 \text{ m/s} + v_{2f} \tag{2}$$

Substituting Equation (1) into Equation (2) and simplifying gives

$$\left(1 + 9.66 \times 10^{-2}\right)v_{1f} = 9.70 \text{ m/s} + \left(\frac{533}{590}\right)(4.85 \text{ m/s})$$

or

$$v_{1f} = +12.84 \text{ m/s}$$

The vertical displacement of the tennis ball during its rebound is then found from $v_y^2 = v_{0y}^2 + 2a_y\Delta y$ to be

$$\Delta y = \frac{v_y^2 - v_{0y}^2}{2a_y} = \frac{0 - (12.84 \text{ m/s})^2}{2\left(-9.80 \text{ m/s}^2\right)} = 8.41 \text{ m} \qquad \Diamond$$

7

Rotational Motion and the Law of Gravity

NOTES FROM SELECTED CHAPTER SECTIONS

7.1 Angular Speed and Angular Acceleration

Pure rotational motion refers to the motion of a rigid body about a fixed axis. In the case of **rotation about a fixed axis,** every particle on the rigid body has the same angular velocity and the same angular acceleration.

One **radian** (rad) is the angle subtended by an arc length equal to the radius of the arc. That is, the angle measured in radians is given by the arc length divided by the corresponding radius. One complete rotation of 360 degrees equals 2π radians.

7.2 Rotational Motion Under Constant Angular Acceleration

The equations for rotational motion under constant angular acceleration are of the same form as those for linear motion under constant linear acceleration with the substitutions $x \to \theta$, $v \to \omega$, and $a \to \alpha$.

7.3 Relations Between Angular and Linear Quantities

When a **rigid body rotates about a fixed axis,** every point in the object moves along a circular path which has its center at the axis of rotation. The instantaneous velocity of each point is directed along a tangent to the circle. Every point on the object experiences the same angular speed; however, points that are different distances from the axis of rotation have different tangential speeds. The value of each of the linear quantities—displacement (s), velocity (v_t), and acceleration (a_t)—is equal to the radial distance from the axis multiplied by the corresponding angular quantity, θ, ω, and α.

7.4 Centripetal Acceleration

In circular motion, the centripetal acceleration is directed inward toward the center of the circle and has a magnitude given either by v^2/r or $r\omega^2$. The force that causes centripetal acceleration acts toward the center of the circular path along which the object moves. If the force vanishes, the object does not continue to move in its circular path; instead, it moves along a straight-line path tangent to the circle.

7.5 Newtonion Gravitation
There are several important features of the law of universal gravitation.

The gravitational force

- is an action-at-a-distance force that always exists between two particles regardless of the medium that separates them,

- varies as the inverse square of the distance between the particles and therefore decreases rapidly with increasing separation, and

- is proportional to the product of their masses.

The gravitational force exerted by a uniform spherical mass on a particle outside the sphere is the same as if the entire mass of the sphere were concentrated at the center.

7.6 Kepler's Laws
Kepler's laws applied to the solar system are:

1. All planets move in elliptical orbits with the Sun at one of the focal points.

2. A line drawn from the Sun to any planet sweeps out equal areas in equal time intervals.

3. The square of the orbital period of any planet is proportional to the cube of the average distance from the planet to the Sun.

EQUATIONS AND CONCEPTS

Arc length, s, is the distance traveled by a particle as it moves along a circular path of radius r. The radial line from the center of the path to the particle sweeps out an angle, θ.

$$\theta = \frac{s}{r} \tag{7.1}$$

$$\theta \, (\text{rad}) = \left(\frac{\pi}{180°} \right) \theta (\text{deg})$$

The **radian,** a unit of angular measure, is the ratio of two lengths (arc length to radius) and hence is a dimensionless quantity.

$$1 \, \text{rad} = \left(\frac{360°}{2\pi} \right) = 57.3°$$

The **average angular velocity** of a rotating object is the ratio of the angular displacement to the time interval during which the angular displacement occurs.

$$\omega_{av} \equiv \frac{\theta_f - \theta_i}{t_f - t_i} = \frac{\Delta\theta}{\Delta t} \tag{7.3}$$

The **average angular acceleration** of a rotating object is the ratio of change in angular velocity to the time interval during which the change in velocity occurs.

$$\alpha_{av} \equiv \frac{\omega_f - \omega_i}{t_f - t_i} = \frac{\Delta\omega}{\Delta t} \tag{7.5}$$

The **equations of rotational kinematics** describe the motion of a particle or extended body rotating about a fixed axis with *constant acceleration*. Note that the rotational equations, involving the angular variables $\Delta\theta$, ω, and α, have a one-to-one correspondence with the equations of linear motion, involving the variables Δx, v, and a.

$$\omega = \omega_i + \alpha t \tag{7.7}$$
$$v = v_i + at$$
$$\Delta\theta = \omega_i t + \tfrac{1}{2}\alpha t^2 \tag{7.8}$$
$$\Delta x = v_i t + \tfrac{1}{2}at^2$$
$$\omega^2 = \omega_i^2 + 2\alpha\,\Delta\theta \tag{7.9}$$
$$v^2 = v_i^2 + 2a\Delta x$$

The **magnitudes of the tangential velocity** and **tangential acceleration** of a given point on a rotating object are related to the corresponding angular quantities via the radius of the path along which the point moves. The tangential velocity and tangential acceleration are directed along the tangent to the circular path (and therefore perpendicular to the radius from the center of rotation). *Every point on a rotating object has the same value of ω and the same value of α.*

$$v_t = r\omega \tag{7.10}$$
$$a_t = r\alpha \tag{7.11}$$

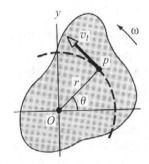

Tangential velocity is perpendicular to the path radius.

The **centripetal acceleration** of an object in circular motion is directed toward the center of the circle (see following figure) with a magnitude that depends on the values of the tangential (or angular) velocity and the radius of the path.

$$a_c = \frac{v^2}{r} \tag{7.13}$$

or

$$a_c = r\omega^2 \tag{7.17}$$

The **total acceleration** of an object in rotational motion has two components as shown in the accompanying diagram.

The **tangential component** (a_t) given by Equation (7.11) is directed along the direction of travel.

$$a = \sqrt{a_t^2 + a_c^2} \tag{7.18}$$

$$\theta = \tan^{-1}\left(\frac{a_t}{a_c}\right)$$

The **centripetal component** (a_c) given by Equation (7.13) is directed perpendicular to the direction of travel.

The **total acceleration** (a) (magnitude and direction) can be found by using Equation (7.18) and the equation for the angle θ.

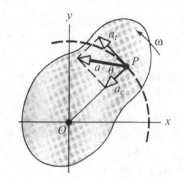

A net **radial (or centripetal)** force is required to maintain motion along a circular path and is always directed toward the center of the path. *Equation (7.19) is a statement of Newton's second law along the radial direction.*

$$F_r = ma_c = m\frac{v^2}{r} \tag{7.19}$$

The **law of universal gravitation** states that every particle in the universe attracts every other particle with a force that is directly proportional to the product of their masses and inversely proportional to the square of the distance between them. The constant G is called the universal gravitational constant. *The magnitudes of the forces of gravitational attraction on the two masses shown in the figure are equal regardless of the relative values of m_1 and m_2.*

$$F = G\frac{m_1 m_2}{r^2} \tag{7.20}$$

$$G = 6.673\times10^{-11}\ \text{N}\cdot\text{m}^2/\text{kg}^2$$

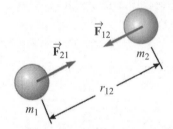

The **escape speed** is the minimum speed that an object projected upward from the Earth's surface must have in order to escape from the Earth's gravitational field. *The escape speed does not depend on the mass of the object.*

$$v_{esc} = \sqrt{\frac{2GM_E}{R_E}} \tag{7.22}$$

Kepler's third law states that the square of the orbital period of a planet is proportional to the cube of the mean distance from the planet to the Sun. The value of K_S is independent of the mass of the planet orbiting the Sun.

$$T^2 = \left(\frac{4\pi^2}{GM_S}\right)r^3 = K_S r^3 \qquad (7.23)$$

$$K_S = \frac{4\pi^2}{GM_S} = 2.97 \times 10^{-19} \text{ s}^2/\text{m}^3$$

For an Earth satellite, (e.g., the moon orbiting the Earth) M_S in Equation (7.23) must be replaced by M_E, the mass of the Earth, and the constant K would have a different value.

SUGGESTIONS, SKILLS, AND STRATEGIES

When solving problems involving rotational motions and centripetal accelerations, the following points should be kept in mind:

1. Draw a free-body diagram of the object(s) under consideration, showing all forces that act on it (them).

2. Choose a coordinate system with one axis along the radial direction (i.e., tangent to the path followed by the object) and the other axis perpendicular to the radial direction.

3. Find the net force toward the center of the circular path. This is the centripetal force, the force which causes the centripetal acceleration.

4. From this point onward, the steps are virtually identical to those encountered when solving Newton's second law problems with $F_x = ma_x$ and $F_y = ma_y$. In this case, Newton's second law is applied along the radial (directed toward the center) and tangential directions. Also, you should note that the magnitude of the centripetal acceleration can always be written for circular motion as $a_c = v^2/r$.

REVIEW CHECKLIST

- Quantitatively, the angular displacement, angular velocity, and angular acceleration for a rigid body system in rotational motion are related to the distance traveled, tangential velocity, and tangential acceleration, respectively. Each linear quantity is calculated by multiplying the corresponding angular quantity by the radius for that particular object or point.

- If a body rotates about a fixed axis, every particle on the body has the same angular velocity and angular acceleration. For this reason, rotational motion can be simply described using these quantities. The formulas which describe angular motion are analogous to the corresponding set of equations pertaining to linear motion.

- A particle moving along a circular path with constant speed experiences an acceleration although the magnitude of the velocity remains constant. This is true because the direction of the velocity (always tangent to the circular path) changes with time. The direction of the acceleration is toward the center of the circular path.

- When both the magnitude and direction of $\vec{v}$ are changing with time, there are two components of acceleration for a particle moving on a curved path. In this case, the particle has a tangential component of acceleration and a radial component of acceleration.

- Newton's law of universal gravitation is an example of an inverse-square law, and it describes an attractive force between two particles separated by a distance r.

SOLUTIONS TO SELECTED END-OF-CHAPTER PROBLEMS

1. (a) Find the angular speed of Earth's rotation about its axis. (b) How does this rotation affect the shape of Earth?

Solution

(a) The rotating Earth makes one full turn $\left(360° \text{ or } 2\pi \text{ radians}\right)$ in a time interval of $\Delta t = 1$ day. The angular speed of this rotation is then

$$\omega_E = \frac{\Delta\theta}{\Delta t} = \frac{2\pi \text{ rad}}{1 \text{ d}}\left(\frac{1 \text{ d}}{24 \text{ h}}\right)\left(\frac{1 \text{ h}}{60 \text{ min}}\right)\left(\frac{1 \text{ min}}{60 \text{ s}}\right) = 7.27 \times 10^{-5} \text{ rad/s} \qquad \lozenge$$

(b) The rotation of Earth about its axis causes it to assume the shape of a slightly flattened sphere, with the radius measured in the equatorial plane being a few miles greater than the radius measured along the rotation axis. When a solid body rotates, all parts of that body have the same angular speed ω. However, even in a perfect spherical shape, the material on the equator of the body is farther from the rotation axis than is material at other locations on the surface. This means that the net force directed in toward the axis, $F_c = mr\omega^2$, needed to produce the centripetal acceleration for material in the equatorial region is greater than for material at other locations. This requirement for a slightly greater centripetal force causes material in the equatorial regions to bulge outward from Earth's center a little farther than does material in the polar regions. $\qquad \lozenge$

9. The diameters of the main rotor and tail rotor of a single-engine helicopter are 7.60 m and 1.02 m, respectively. The respective rotational speeds are 450 rev/min and 4 138 rev/min. Calculate the speeds of the tips of both rotors. Compare these speeds with the speed of sound, 343 m/s.

Solution

The tips of one of the rotors move in a circular path with the specified angular velocity (or rotational speed), ω, for that rotor. The linear speed of these rotor tips is given by $v_t = r\omega$, where r is the radius of the circular path followed by the tips.

For the main rotor:

$$v_t = r\omega = \left(\frac{7.60 \text{ m}}{2}\right)\left(450 \ \frac{\text{rev}}{\text{min}}\right)\left(\frac{1 \text{ min}}{60 \text{ s}}\right)\left(\frac{2\pi \text{ rad}}{1 \text{ rev}}\right) = 179 \text{ m/s} \qquad \Diamond$$

and

$$v_t = (179 \text{ m/s})\left(\frac{v_{sound}}{343 \text{ m/s}}\right) = 0.522 \, v_{sound} \qquad \Diamond$$

For the tail rotor:

$$v_t = r\omega = \left(\frac{1.02 \text{ m}}{2}\right)\left(4\,138 \ \frac{\text{rev}}{\text{min}}\right)\left(\frac{1 \text{ min}}{60 \text{ s}}\right)\left(\frac{2\pi \text{ rad}}{1 \text{ rev}}\right) = 221 \text{ m/s} \qquad \Diamond$$

and

$$v_t = (221 \text{ m/s})\left(\frac{v_{sound}}{343 \text{ m/s}}\right) = 0.644 \, v_{sound} \qquad \Diamond$$

---

13. A rotating wheel requires 3.00-s to rotate 37.0 revolutions. Its angular velocity at the end of the 3.00-s interval is 98.0 rad/s. What is the constant angular acceleration of the wheel?

Solution

For the 3.00-s time interval of interest, we have the following information about the motion of the wheel:

Angular displacement undergone: $\Delta\theta = (37.0 \text{ rev})\left(\frac{2\pi \text{ rad}}{1 \text{ rev}}\right) = 74.0\pi \text{ rad}$

Final angular velocity: $\omega = 98.0 \text{ rad/s}$

Elapsed time: $t = 3.00 \text{ s}$

The unknown quantities about the motion are

Initial angular velocity: ω_0

Constant angular acceleration: α

From the relation $\omega = \omega_0 + \alpha t$, we obtain: $\omega_0 = \omega - \alpha t$. Substituting this into the relation $\Delta\theta = \omega_0 t + \frac{1}{2}\alpha t^2$ yields

$$\Delta\theta = (\omega - \alpha t)t + \frac{1}{2}\alpha t^2 = \omega t - \frac{1}{2}\alpha t^2$$

or

$$\alpha = \frac{2(\omega t - \Delta\theta)}{t^2}$$

Thus, the angular acceleration of the wheel is

$$\alpha = \frac{2\left[(98.0 \ \text{rad/s})(3.00 \ \text{s}) - 74.0\pi \ \text{rad}\right]}{(3.00 \ \text{s})^2} = 13.7 \ \text{rad/s}^2 \qquad \Diamond$$

If desired, the initial angular velocity of the wheel can now be found:

$$\omega_0 = \omega - \alpha t = 98.0 \ \text{rad/s} - (13.7 \ \text{rad/s}^2)(3.00 \ \text{s}) = 57.0 \ \text{rad/s}$$

---

17. (a) What is the tangential acceleration of a bug on the rim of a 10-in.-diameter disk if the disk moves from rest to an angular speed of 78 rev/min in 3.0 s? (b) When the disk is at its final speed, what is the tangential velocity of the bug? (c) One second after the bug starts from rest, what are its tangential acceleration, centripetal acceleration, and total acceleration?

Solution

The angular velocity of the disk 3.0 s after starting from rest is

$$\omega = 78 \ \frac{\text{rev}}{\text{min}} \left(\frac{2\pi \ \text{rad}}{1 \ \text{rev}}\right)\left(\frac{1 \ \text{min}}{60 \ \text{s}}\right) = 8.2 \ \text{rad/s}$$

and the bug follows a circular path of radius

$$r = 5.0 \ \text{in}\left(\frac{1 \ \text{m}}{39.37 \ \text{in}}\right) = 0.13 \ \text{m}$$

(a) The constant angular acceleration of the disk is

$$\alpha = \frac{\omega - \omega_i}{t} = \frac{8.2 \text{ rad/s} - 0}{3.0 \text{ s}} = 2.7 \text{ rad/s}^2$$

The tangential acceleration of the bug is $a_t = r\alpha$. Thus,

$$a_t = (0.13 \text{ m})(2.7 \text{ rad/s}^2) = 0.35 \text{ m/s}^2 \qquad \lozenge$$

(b) When the disk is rotating at its final angular velocity, $\omega = 8.2 \text{ rad/s}$, and

$$v_t = r\omega = (0.13 \text{ m})(8.2 \text{ rad/s}) = 1.0 \text{ m/s} \qquad \lozenge$$

(c) Since both r and α are constant, the tangential acceleration, $a_t = r\alpha$, is also constant. Thus, at $t = 1.0$ s, $a_t = 0.35 \text{ m/s}^2$. $\qquad \lozenge$

At $t = 1.0$ s, the tangential velocity of the bug is

$$v_t = (v_t)_{t=0} + a_t t = 0 + (0.35 \text{ m/s}^2)(1.0 \text{ s}) = 0.35 \text{ m/s}$$

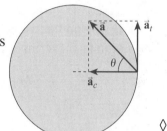

and the centripetal acceleration is

$$a_c = \frac{v_t^2}{r} = \frac{(0.35 \text{ m/s})^2}{0.13 \text{ m}} = 0.94 \text{ m/s}^2 \qquad \lozenge$$

At $t = 1.0$ s, the total acceleration has magnitude

$$a = \sqrt{a_c^2 + a_t^2} = \sqrt{(0.94 \text{ m/s}^2)^2 + (0.35 \text{ m/s}^2)^2} = 1.0 \text{ m/s}^2 \qquad \lozenge$$

at

$$\theta = \tan^{-1}\left(\frac{a_t}{a_c}\right) = \tan^{-1}\left(\frac{0.35 \text{ m/s}^2}{0.94 \text{ m/s}^2}\right) = 20° \qquad \lozenge$$

27. An air puck of mass 0.25 kg is tied to a string and allowed to revolve in a circle of radius 1.0 m on a frictionless horizontal table. The other end of the string passes through a hole in the center of the table, and a mass of 1.0 kg is tied to it (Fig. P7.27). The suspended mass remains in equilibrium while the puck on the tabletop revolves. (a) What is the tension in the string? (b) What is the horizontal force acting on the puck? (c) What is the speed of the puck?

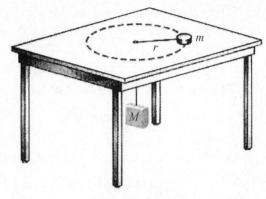

Figure P7.27

Solution

(a) Since the suspended mass on the end of the string is in equilibrium, the resultant force acting on it must be zero. The forces acting on this object are the upward directed tension in the string and the downward gravitational force. Therefore,

$$\Sigma F_y = T - Mg = 0$$

or

$$T = Mg = (1.0 \text{ kg})(9.80 \text{ m/s}^2) = 9.8 \text{ N} \qquad \Diamond$$

(b) The horizontal force acting on the puck is the tension in the portion of the string that is above the table. Assuming that the string is free to slide, without friction, through the hole in the tabletop, the tension is uniform throughout the length of the string. In this case the magnitude of the horizontal force acting on the puck is $T = 9.8$ N as found in Part (a). $\qquad \Diamond$

(c) The puck, moving in a circular path of radius r, has a centripetal acceleration $a_c = v^2/r$. The net force acting on the puck and directed toward the center of the circular path (the tension in the string in this case) must produce this acceleration. Newton's second law then gives $F_c = T = mv^2/r$, or

$$v = \sqrt{\frac{Tr}{m}} = \sqrt{\frac{(9.8 \text{ N})(1.0 \text{ m})}{0.25 \text{ kg}}} = 6.3 \text{ m/s} \qquad \Diamond$$

36. After the Sun exhausts its nuclear fuel, its ultimate fate may be to collapse to a *white dwarf* state. In this state, it would have approximately the same mass as it has now, but its radius would be equal to the radius of Earth. Calculate (a) the average density of the white dwarf, (b) the surface free-fall acceleration, and (c) the gravitational potential energy associated with a 1.00-kg object at the surface of the white dwarf.

Solution

(a) The average density of an object is the ratio of the mass of the object to its volume (i.e., the mass per unit volume.) For a spherical object, this is

$$\rho_{av} = \frac{M}{V} = \frac{M}{4\pi r^3/3} = \frac{3M}{4\pi r^3}$$

When the Sun collapses to a white dwarf state, its mass will be $M \approx M_S = 1.991 \times 10^{30}$ kg and its radius will be $r \approx R_E = 6.38 \times 10^6$ m. (Note that these values and similar data on some members of our solar system are given in Table 7.3 of the textbook.) The average density of the white dwarf will be

$$\rho_{av} = \frac{3(1.991 \times 10^{30} \text{ kg})}{4\pi(6.38 \times 10^6 \text{ m})^3} = 1.83 \times 10^9 \text{ kg/m}^3 \qquad \Diamond$$

(b) Newton's law of universal gravitation gives the gravitational force acting on an object of mass m on the surface of the white dwarf as $F_g = mg = GM_S m/R_E^2$. Therefore, the free-fall acceleration on the surface is

$$g = \frac{F_g}{m} = \frac{GM}{R_E^2} = \frac{(6.67 \times 10^{-11} \text{ N} \cdot \text{m}^2/\text{kg}^2)(1.991 \times 10^{30} \text{ kg})}{(6.38 \times 10^6 \text{ m/s})^2} = 3.26 \times 10^6 \text{ m/s}^2 \qquad \Diamond$$

(c) Outside of a spherical distribution of matter having mass M, the gravitational potential energy of an object of mass m located distance r from the center is $PE_g = -GMm/r$. The gravitational potential energy of a 1.00-kg mass on the surface of the white dwarf is then

$$PE_g = -\frac{GM_S m}{R_E} = -\frac{(6.67 \times 10^{-11} \text{ N} \cdot \text{m}^2/\text{kg}^2)(1.991 \times 10^{30} \text{ kg})(1.00 \text{ kg})}{6.38 \times 10^6 \text{ m}}$$

or

$$PE_g = -2.08 \times 10^{13} \text{ J} \qquad \Diamond$$

45. A satellite of mass 200 kg is launched from a site on Earth's equator into an orbit 200 km above the surface of Earth. (a) Assuming a circular orbit, what is the orbital period of this satellite? (b) What is the satellite's speed in its orbit? (c) What is the minimum energy necessary to place the satellite in orbit, assuming no air friction?

Solution

The radius of the satellite's orbit is

$$r = R_E + 200 \text{ km} = 6.38 \times 10^6 \text{ m} + 200 \times 10^3 \text{ m} = 6.58 \times 10^6 \text{ m}$$

(a) The required centripetal acceleration of the satellite is produced by the gravitational force exerted on it by Earth.

Thus,

$$G \frac{M_E m}{r^2} = m \left(\frac{v_t^2}{r} \right)$$

which yields

$$v_t = \sqrt{\frac{GM_E}{r}}$$

The orbital speed of the satellite is therefore

$$v_t = \sqrt{\frac{\left(6.67 \times 10^{-11} \text{ N} \cdot \text{m}^2/\text{kg}^2\right)\left(5.98 \times 10^{24} \text{ kg}\right)}{6.58 \times 10^6 \text{ m}}} = 7.79 \times 10^3 \text{ m/s}$$

and its period is

$$T = \frac{\text{circumference}}{\text{speed}} = \frac{2\pi r}{v_t} = \frac{2\pi\left(6.58 \times 10^6 \text{ m}\right)}{7.79 \times 10^3 \text{ m/s}} = 5.31 \times 10^3 \text{ s} = 1.48 \text{ h} \qquad \Diamond$$

(b) The orbital speed of the satellite was found above to be

$$v_t = 7.79 \times 10^3 \text{ m/s} \qquad \Diamond$$

(c) When the satellite is in orbit, its gravitational potential energy is

$$\left(PE_g\right)_f = -G \frac{M_E m}{r} = -\left(6.67 \times 10^{-11} \frac{\text{N} \cdot \text{m}^2}{\text{kg}^2}\right) \frac{\left(5.98 \times 10^{24} \text{ kg}\right)\left(200 \text{ kg}\right)}{6.58 \times 10^6 \text{ m}}$$

$$= -1.21 \times 10^{10} \text{ J}$$

and its kinetic energy is

$$KE_f = \frac{1}{2} m v_t^2 = \frac{1}{2}\left(200 \text{ kg}\right)\left(7.79 \times 10^3 \text{ m/s}\right)^2 = 6.06 \times 10^9 \text{ J}$$

Initially the satellite was moving with the rotational speed of a point on the equator of Earth, or

$$v_i = \frac{\Delta s}{\Delta t} = \frac{2\pi R_E}{24.0 \text{ h}} = \frac{2\pi (6.38 \times 10^6 \text{ m})}{24.0 \cancel{h}} \left(\frac{1 \cancel{h}}{3\,600 \text{ s}} \right) = 464 \text{ m/s}$$

Its initial kinetic energy was

$$KE_i = \frac{1}{2} m v_i^2 = \frac{1}{2} (200 \text{ kg})(464 \text{ m/s})^2 = 2.15 \times 10^7 \text{ J}$$

and its gravitational potential energy was $\left(PE_g\right)_i = -G\dfrac{M_E m}{R_e}$, or

$$\left(PE_g\right)_i = -\left(6.67 \times 10^{-11} \ \frac{\text{N} \cdot \text{m}^2}{\text{kg}^2} \right) \frac{(5.98 \times 10^{24} \text{ kg})(200 \text{ kg})}{6.38 \times 10^6 \text{ m}} = -1.25 \times 10^{10} \text{ J}$$

The work-energy theorem then gives the energy required to place the satellite in orbit as $W_{nc} = \left(KE + PE\right)_f - \left(KE + PE\right)_i$

or

$$W_{nc} = \left(6.06 \times 10^9 \text{ J} - 1.21 \times 10^{10} \text{ J}\right) - \left(2.15 \times 10^7 \text{ J} - 1.25 \times 10^{10} \text{ J}\right)$$

$$W_{nc} = 6.43 \times 10^9 \text{ J} \qquad\qquad\qquad \Diamond$$

---

49. One method of pitching a softball is called the "windmill" delivery method, in which the pitcher's arm rotates through approximately 360° in a vertical plane before the 198-gram ball is released at the lowest point of the circular motion. An experienced pitcher can throw a ball with a speed of 98.0 mi/h. Assume that the angular acceleration is uniform throughout the pitching motion and take the distance between the softball and the shoulder joint to be 74.2 cm. (a) Determine the angular speed of the arm in rev/s at the instant of release. (b) Find the value of the angular acceleration in rev/s² and the radial and tangential acceleration of the ball just before it is released. (c) Determine the force exerted on the ball by the pitcher's hand (both radial and tangential components) just before it is released.

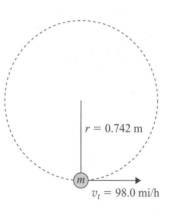

Solution

(a) The ball, at the end of the arm, has a tangential speed of $v_t = 98.0$ mi/h just before it is released at the lowest point on the circular path. At this instant, the ball, hand, and arm have an angular speed of

$$\omega = \frac{v_t}{r} = \frac{98.0 \text{ mi/h}}{0.742 \text{ m}} \left(\frac{1 \text{ m/s}}{2.237 \text{ mi/h}} \right) \left(\frac{1 \text{ rev}}{2\pi \text{ rad}} \right) = 9.40 \text{ rev/s} \qquad \Diamond$$

(b) Assuming the ball has had a uniform angular acceleration during the entire 360° or 1 revolution of the pitching motion, the angular acceleration is given by the rotational kinematics relation $\omega^2 = \omega_i^2 + 2\alpha(\Delta\theta)$ as

$$\alpha = \frac{\omega^2 - \omega_i^2}{2(\Delta\theta)} = \frac{(9.40 \text{ rev/s})^2 - 0}{2(1 \text{ rev})} = 44.2 \text{ rev/s}^2 \qquad \Diamond$$

The tangential acceleration of the ball is $a_t = r\alpha$, where α is expressed in units of rad/s². Therefore,

$$a_t = r\alpha = (0.742 \text{ m})\left[\left(44.1 \frac{\text{rev}}{\text{s}^2}\right)\left(\frac{2\pi \text{ rad}}{1 \text{ rev}}\right)\right] = 206 \text{ m/s}^2 \qquad \Diamond$$

The radial, or centripetal, acceleration of the ball at the moment of release is

$$a_c = \frac{v_t^2}{r} = \frac{\left[(98.0 \text{ mi/h})\left(\dfrac{1 \text{ m/s}}{2.237 \text{ mi/h}}\right)\right]^2}{0.742 \text{ m}} = 2.59 \times 10^3 \text{ m/s}^2 \qquad \Diamond$$

(c) The sketch at the right is a free-body diagram of the ball at the moment of release. The net force directed toward the center of the circular path is the force that produces the centripetal acceleration. From Newton's second law, $F_c = F_r - mg = ma_c$. The radial component of force exerted on the ball by the arm is then

$$F_r = ma_c + mg = m(a_c + g) = (0.198 \text{ kg})(2.59 \times 10^3 \text{ m/s}^2 + 9.80 \text{ m/s}^2) = 515 \text{ N} \qquad \Diamond$$

The tangential component of force the arm exerts on the ball gives the ball its tangential acceleration. This component is given by

$$F_t = ma_t = (0.198 \text{ m/s}^2)(206 \text{ m/s}^2) = 40.8 \text{ N} \qquad \Diamond$$

57. Because of Earth's rotation about its axis, a point on the equator experiences a centripetal acceleration of $0.034\ 0\ \text{m/s}^2$, while a point at the poles has no centripetal acceleration. (a) Show that, at the equator, the gravitational force on an object (the object's true weight) must exceed the object's apparent weight. (b) What are the apparent weights of a 75.0-kg person at the equator and at the poles? (Assume Earth is a uniform sphere, and take $g = 9.800\ \text{m/s}^2$.)

Solution

(a) The sketch at the right is a view from above the north pole of Earth, showing an object on the equator being weighed with a spring scale. Note that this object is moving in a circular path of radius $r = R_E$ with the angular speed of the rotating Earth. The scale reading (the apparent weight of the object) is equal to the upward normal force exerted on the object by the scale. The true weight of the object is equal to the downward gravitational force exerted on the object by Earth.

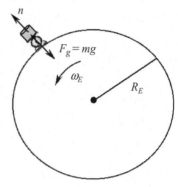

The net force directed toward the center of the circular path must supply the centripetal acceleration. Thus, $\Sigma F_{\text{toward center}} = F_g - n = ma_c$ and the true weight is

$$F_g = n + ma_c = apparent\ weight + ma_c > apparent\ weight \qquad \Diamond$$

(b) From above, the apparent weight of an object is $n = F_g - ma_c = m\,(g - a_c)$. The centripetal acceleration of an object on the rotating Earth is $a_c = r\omega_E^2$, where r is the radius of the circular path the object follows around Earth's axis.

At the equator, $r = R_E$, and it is given that $a_c = R_E\omega_E^2 = 0.034\ 0\ \text{m/s}^2$. The apparent weight of a 75.0-kg person at the equator is

$$n = m(g - a_c) = (75.0\ \text{kg})(9.800\ \text{m/s}^2 - 0.034\ 0\ \text{m/s}^2) = 732\ \text{N} \qquad \Diamond$$

At the north pole, the radius of the circular path followed by the person is $r = 0$ and the centripetal acceleration is $a_c = r\omega_E^2 = 0$. The apparent weight of the person here is

$$n = m(g - a_c) = (75.0\ \text{kg})(9.800\ \text{m/s}^2 - 0) = 735\ \text{N} = \text{the true weight} \qquad \Diamond$$

61. Assume that you are agile enough to run across a horizontal surface at 8.50 m/s, independently of the value of the gravitational field. What would be (a) the radius and (b) the mass of an airless spherical asteroid of uniform density 1.10×10^3 kg/m^3 on which you could launch yourself into orbit by running? (c) What would be your period?

Solution

(a) If you are to launch yourself into orbit by running, the minimum running speed is that for which the gravitational force acting on you exactly matches the force needed to produce the required centripetal acceleration. That is, we must have

$$G \frac{Mm}{R^2} = m \left(\frac{v_t^2}{R} \right)$$

where M is the mass of the asteroid and R is its radius. Since the mass of the asteroid may be written as

$$M = \text{density} \times \text{volume} = \rho \left(\frac{4}{3} \pi R^3 \right)$$

this requirement becomes

$$G \rho \left(\frac{4}{3} \pi R^3 \right) \frac{\not{m}}{R^2} = \not{m} \left(\frac{v_t^2}{R} \right)$$

or

$$R = \sqrt{\frac{3 v_t^2}{4 \pi G \rho}}$$

The radius of the asteroid would then be

$$R = \sqrt{\frac{3(8.50 \text{ m/s})^2}{4\pi \left(6.67 \times 10^{-11} \text{ N} \cdot \text{m}^2/\text{kg}^2\right)\left(1.10 \times 10^3 \text{ kg/m}^3\right)}}$$

or

$$R = 1.53 \times 10^4 \text{ m} = 15.3 \text{ km} \qquad \lozenge$$

(b) The mass of the asteroid will be

$$M = \rho \left(\frac{4}{3} \pi R^3 \right) = \frac{4}{3} \left(1.10 \times 10^3 \text{ kg/m}^3\right) \pi \left(1.53 \times 10^4 \text{ m}\right)^3 = 1.65 \times 10^{16} \text{ kg} \qquad \lozenge$$

(c) Your orbital period would be

$$T = \frac{2\pi}{\omega} = \frac{2\pi R}{v_t} = \frac{2\pi \left(1.53 \times 10^4 \text{ m}\right)}{8.50 \text{ m/s}} = 1.13 \times 10^4 \text{ s} = 3.14 \text{ h} \qquad \lozenge$$

70. A 0.275-kg object is swung in a vertical circular path on a string 0.850 m long as in Figure P7.70. (a) What are the forces acting on the ball at any point along this path? (b) Draw free-body diagrams for the ball when it is at the bottom of the circle and when it is at the top. (c) If its speed is 5.20 m/s at the top of the circle, what is the tension in the string there? (d) If the string breaks when its tension exceeds 22.5 N, what is the maximum speed the object can have at the bottom before the string breaks?

Solution

(a) At any point on the circular path, the ball has two forces acting on it. One is the gravitational force, with the constant magnitude $F_g = mg$, and is always directed vertically downward. The other is the tension in the string, which is always directed inward toward the center of the circular path but varies in magnitude from point to point on the path. ◊

(b) The sketch at the right includes two free-body diagrams, one of the ball when at the bottom of the circular path, and one when the ball is at the highest point on the path. Note how the tension force is directed toward the center in both cases, while the gravitational force is always directed downward. The vector sum of these forces must provide the needed centripetal acceleration at each point on the path.

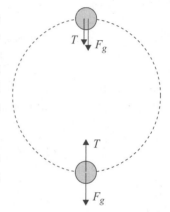

(c) At the top of the circle, the net force directed toward the center is $F_c = T + mg = mv^2/r$. The tension in the string at this point on the path must be

$$T = m\left(\frac{v^2}{r} - g\right) = (0.275 \text{ kg})\left(\frac{(5.20 \text{ m/s})^2}{0.850 \text{ m}} - 9.80 \text{ m/s}^2\right) = 6.05 \text{ N} \qquad ◊$$

(d) At the lowest point on the circular path, the net force toward the center is $F_c = T - mg = mv^2/r$. Thus, if the tension in the string at this point is to be $T = T_{max}$, the speed of the ball must be $v = \sqrt{(r/m)(T_{max} - mg)}$. With $T_{max} = 22.5$ N, this speed is

$$v = \sqrt{\left(\frac{0.850 \text{ m}}{0.275 \text{ kg}}\right)\left[22.5 \text{ N} - (0.275 \text{ kg})(9.80 \text{ m/s}^2)\right]} = 7.82 \text{ m/s} \qquad ◊$$

75. In a popular amusement park ride, a rotating cylinder of radius 3.00 m is set in rotation at an angular speed of 5.00 rad/s, as in Figure P7.75. The floor then drops away, leaving the riders suspended against the wall in a vertical position. What minimum coefficient of friction between a rider's clothing and the wall is needed to keep the rider from slipping? (*Hint*: Recall that the magnitude of the maximum force of static friction is equal to $\mu_s n$, where n is the normal force—in this case, the force causing the centripetal acceleration.)

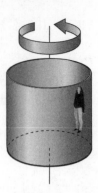

Figure P7.75

Solution

The normal force exerted on the person by the cylindrical wall must provide the centripetal acceleration, so

$$n = m\left(\frac{v_t^2}{r}\right) = \frac{m(r\omega)^2}{r} = m(r\omega^2)$$

If the minimum acceptable coefficient of friction is present, the person is on the verge of slipping and the maximum static friction force equals the person's weight, or

$$(f_s)_{max} = (\mu_s)_{min}\, n = mg$$

Thus,
$$(\mu_s)_{min} = \frac{mg}{n} = \frac{g}{r\omega^2} = \frac{9.80 \text{ m/s}^2}{(3.00 \text{ m})(5.00 \text{ rad/s})^2} = 0.131 \qquad \Diamond$$

8

Rotational Equilibrium and Rotational Dynamics

NOTES FROM SELECTED CHAPTER SECTIONS

8.1 Torque

Torque is the physical quantity that is a measure of the tendency of a force to cause rotation of a body about a specified axis. **Torque must be defined with respect to a specific axis of rotation.** Torque, which has the SI **units** of N·m, must not be confused with work and energy.

8.2 Torque and the Two Conditions For Equilibrium

A body in static equilibrium must satisfy two conditions:

1. The resultant external force must be zero.

2. The resultant external torque about any axis must be zero.

8.3 The Center of Gravity

In order to calculate the torque due to the weight (gravitational force) on a rigid body, the entire weight of the object can be considered to be concentrated at a single point called the center of gravity. *The center of gravity of a homogeneous, symmetric body must lie along an axis of symmetry*

8.5 Relationship Between Torque and Angular Acceleration

The angular acceleration of an object is proportional to the net torque acting on it. The moment of inertia of the object is the proportionality constant between the net torque and the angular acceleration. **The force and mass in linear motion correspond to torque and moment of inertia in rotational motion.** Moment of inertia of an object depends on the location of the axis of rotation and upon the manner in which the mass is distributed relative to that axis (e.g., a ring has a greater moment of inertia than a disk of the same mass and radius).

8.6 Rotational Kinetic Energy

In linear motion, the energy concept is useful in describing the motion of a system. The energy concept can be equally useful in simplifying the analysis of rotational motion. We now have expressions for four types of mechanical energy: gravitational potential energy, PE_g; elastic potential energy, PE_s; translational kinetic energy, KE_t; and rotational kinetic energy, KE_r. We must include all these forms of energy in the equation for conservation of mechanical energy.

EQUATIONS AND CONCEPTS

The **magnitude of the torque** about a specified axis due to a given force depends on

- magnitude of the force

- point of application of the force

- direction of the force

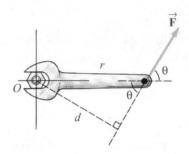

The **lever arm** is the perpendicular distance from the axis of rotation to the line along which the force is acting.

$$\text{lever arm} = d = r \sin\theta$$

$$\tau = Fr \sin\theta \qquad (8.2)$$

$$\tau = d \sin\theta$$

The **algebraic sign of a torque** due to a force is considered positive if the force has a tendency to rotate the body counterclockwise about the chosen axis, and negative if the tendency for rotation is clockwise. In the figure at right with axis perpendicular to the page at point O:
The torque due to $\vec{F}_1$ is positive.
The torque due to $\vec{F}_2$ is negative.

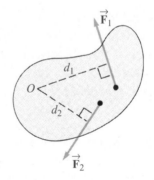

The **first condition for equilibrium** requires that the net external force acting on a body equal zero.

$$\Sigma\vec{F} = 0$$

Translational equilibrium

The **second condition for equilibrium** requires that the net external torque acting on a body equals zero.

$$\Sigma\vec{\tau} = 0$$

Rotational equilibrium

The **center of gravity** is the point where the total weight of an object can be considered concentrated when computing the torque due to the force of gravity (an object's weight). The coordinates of the center of gravity are x_{cg}, y_{cg}, and z_{cg}. *The center of gravity of a symmetric homogeneous body must lie on its axis of symmetry but not necessarily on the object.*

$$x_{cg} = \frac{\Sigma m_i x_i}{\Sigma m_i} \qquad (8.3a)$$

$$y_{cg} = \frac{\Sigma m_i y_i}{\Sigma m_i} \qquad (8.3b)$$

$$z_{cg} = \frac{\Sigma m_i z_i}{m_i} \qquad (8.3c)$$

The **angular acceleration** of a point mass, moving in a path of radius r, is proportional to the net torque acting on the mass.

$$\tau = (mr^2)\alpha \qquad (8.5)$$

The Equation for calculating moment of inertia of

(1) A point mass m, at a radial distance r, from a specified axis of rotation is shown in the near figure.

(2) A collection of discrete point masses, each with its corresponding values of m and r is shown in the far figure.

(3) Extended rigid bodies with a high degree of symmetry are shown in Table 8.1 of your textbook.

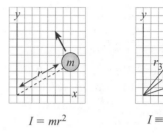

$$I = mr^2 \qquad I \equiv \sum_i m_i r_i^2$$

In all cases, the moment of inertia depends on the mass and distribution of the mass relative to the axis of rotation.

The **angular acceleration** of an extended object is proportional to the net torque acting on the object. *Equation (8.8) is the rotational analog of Newton's second law.*

$$\Sigma \tau = I\alpha \qquad (8.8)$$

Rotational kinetic energy is associated with a rigid body in rotational motion. *Note that this is not a new form of energy but is a convenient form for representing kinetic energy associated with rotational motion.*

$$KE_r \equiv \tfrac{1}{2}I\omega^2 \qquad (8.10)$$

When only conservative forces act on a system, the total mechanical energy of the system (the sum of the gravitational, the elastic potential, the rotational kinetic, and translational kinetic energies) is conserved.

$$(KE_t + KE_r + PE)_i = (KE_t + KE_r + PE)_f \qquad (8.11)$$

When non-conservative forces act on a system, the total mechanical energy of the system changes by an amount equal to the work done by the non-conservative forces.

$$W_{nc} = \Delta KE_t + \Delta KE_r + \Delta PE \qquad (8.12)$$

Angular momentum is a vector quantity associated with rotational motion. The magnitude of angular momentum is defined in Equation (8.13) and must be determined relative to a specified axis. *As shown in the figure, the direction of angular momentum is that of angular velocity.*

$$L \equiv I\omega \qquad (8.13)$$

The **net external torque** acting on an object equals the time rate of change of its angular momentum. *Equation (8.14) is the rotational analog of Newton's second law for transitional motion.*

$$\Sigma\tau = \frac{\Delta L}{\Delta t} \qquad (8.14)$$

Conservation of angular momentum applies when the net external torque acting on a system is zero.

$$L_i = L_f \qquad \text{if} \qquad \Sigma\tau = 0 \qquad (8.15)$$

$$I_i\omega_i = I_f\omega_f \qquad \text{if} \qquad \Sigma\tau = 0 \qquad (8.16)$$

SUGGESTIONS, SKILLS, AND STRATEGIES

PROBLEM-SOLVING STRATEGY FOR OBJECTS IN EQUILIBRIUM

1. Draw a simple, neat diagram of the system that is large enough to show all the forces clearly.

2. Isolate the object of interest being analyzed. Draw a free-body diagram for this object showing all external forces acting on the object. For systems containing more than one object, draw separate diagrams for each object. Do not include forces that the object exerts on its surroundings.

3. Establish convenient coordinate axes for each body and find the components of the forces along these axes. Now apply the first condition of equilibrium for each object under consideration; namely, that the net force on the object in the x- and y-directions must be zero.

4. Choose a convenient origin (axis of rotation) for calculating the net torque on the object. Now apply the second condition of equilibrium that says that the net torque on the object about any origin must be zero. Remember that the choice of the origin for the torque equation is arbitrary; therefore, choose an origin that will simplify your calculation as much as possible. Note that a force that acts along a line passing through the point chosen as the axis of rotation gives zero contribution to the torque because the lever arm is zero in this case.

5. The first and second conditions for equilibrium will give a set of simultaneous equations with several unknowns. To complete your solution, all that is left is to solve for the unknowns in terms of the known quantities.

PROBLEM-SOLVING STRATEGY FOR ROTATIONAL MOTION

The following facts and procedures should be kept in mind when solving rotational motion problems.

1. Problems involving the equation $\Sigma\vec{\tau} = I\vec{\alpha}$ are very similar to those encountered in Newton's second law problems, $\Sigma\vec{F} = m\vec{a}$. Note the correspondences between linear and rotational quantities in that $\vec{F}$ is replaced by $\vec{\tau}$, m by I, and $\vec{a}$ by $\vec{\alpha}$.

2. Other analogues between rotational quantities and linear quantities include the replacement of x by θ and v by ω. Recall that each linear quantity (x, v, and a) equals the product of the radius and the corresponding angular quantity (θ, ω, and α). These are helpful as memory devices for such rotational motion quantities as rotational kinetic energy, $KE_r = \frac{1}{2}I\omega^2$, and angular momentum, $L = I\omega$.

3. With the analogues mentioned in Step 2, conservation of energy techniques remain the same as those examined in Chapter 5, except now an additional term representing rotational kinetic energy must be included in the expression for the conservation of energy. See Equation (8.11).

4. Likewise, the techniques for solving conservation of angular momentum problems are essentially the same as those used in solving conservation of linear momentum problems, except you are equating total angular momentum before to total angular momentum after as $I_i\omega_i = I_f\omega_f$.

REVIEW CHECKLIST

- There are two necessary conditions for equilibrium of a rigid body: $\Sigma F = 0$ and $\Sigma\tau = 0$. Torques that cause counterclockwise rotations are positive and those causing clockwise rotations are negative.

- The torque associated with a force has a magnitude equal to the force times the lever arm. The lever arm is the perpendicular distance from the axis of rotation to a line drawn along the direction of the force. Also, the net torque on a rigid body about some axis is proportional to the angular acceleration; i.e., $\Sigma\tau = I\alpha$, where I is the moment of inertia about the axis for which the net torque is evaluated.

- The work-energy theorem can be applied to a rotating rigid body. That is, the net work done on a rigid body rotating about a fixed axis equals the change in its rotational kinetic energy. The law of conservation of mechanical energy can be used in the solution of problems involving rotating rigid bodies.

- The time rate of change of the angular momentum of a rigid body rotating about an axis is proportional to the net torque acting about the axis of rotation. This is the rotational analog of Newton's second law.

SOLUTIONS TO SELECTED END-OF-CHAPTER PROBLEMS

7. The arm in Figure P8.7 weighs 41.5 N. The force of gravity acting on the arm acts through point A. Determine the magnitudes of the tension force $\vec{F}_t$ in the deltoid muscle and the force $\vec{F}_s$ exerted by the shoulder on the humerus (upper-arm bone) to hold the arm in the position shown.

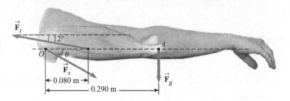

Figure P8.7

Solution

For the easiest solution of this problem, one should resolve the unknown forces $\vec{F}_t$ and $\vec{F}_s$ into their horizontal and vertical components as shown in the free-body diagram at the right. Then, it is observed that the lines of action of many of the unknown components pass through point O, meaning that these components have zero torque about a pivot at point O. Thus, choosing point O as the pivot when applying the second condition for equilibrium, $\Sigma\vec{\tau} = 0$, will lead to a simple equation to solve:

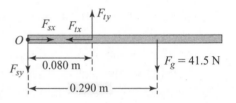

$$\Sigma\vec{\tau}_O = 0 \quad \Rightarrow \quad F_{ty}(0.080 \text{ m}) - (41.5 \text{ N})(0.290 \text{ m}) = 0$$

and since $F_{ty} = F_t \sin 12.0°$, the magnitude of the tension force is

$$F_t = \frac{(41.5 \text{ N})(0.290 \text{ m})}{(0.080 \text{ m})\sin 12.0°} = 724 \text{ N} \qquad \Diamond$$

Then

$$\Sigma F_x = 0 \quad \Rightarrow \quad F_{sx} = F_{tx} = F_t \cos 12.0° = (724 \text{ N})\cos 12.0° = 708 \text{ N}$$

and

$$\Sigma F_y = 0 \quad \Rightarrow \quad F_{sy} = F_{ty} - F_g = (724 \text{ N})\sin 12.0° - 41.5 \text{ N} = 109 \text{ N}$$

so the magnitude of the shoulder force is

$$F_s = \sqrt{F_{sx}^2 + F_{sy}^2} = \sqrt{(708 \text{ N})^2 + (109 \text{ N})^2} = 716 \text{ N} \qquad \Diamond$$

Then, if desired,

$$\tan\theta = \frac{F_s \sin\theta}{F_s \cos\theta} = \frac{F_{sy}}{F_{sx}}$$

or

$$\theta = \tan^{-1}\left(\frac{F_{sy}}{F_{sx}}\right) = \tan^{-1}\left(\frac{109 \text{ N}}{708 \text{ N}}\right) = 8.75°$$

11. Find the x- and y-coordinates of the center of gravity of a 4.00-ft by 8.00-ft uniform sheet of plywood with the upper right quadrant removed as shown in Figure P8.11.

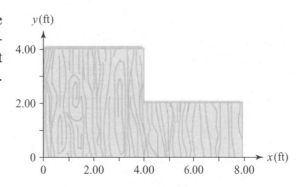

Figure P8.11

Solution

The location of the center of gravity of a uniform body is located at the geometric center of that body. While we cannot easily recognize the geometric center of the irregular shaped sheet of plywood in Figure P8.11, we can think of this sheet as consisting of two areas A_1 and A_2 as shown in the sketch at the right.

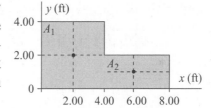

The geometric center (and hence center of gravity) of each of these areas can be found by inspection. The center of gravity of A_1 is located at ($x_1 = 2.00$ ft, $y_1 = 2.00$ ft) and that of A_2 is at ($x_2 = 6.00$ ft, $y_2 = 1.00$ ft). The mass of each portion of plywood is $m_1 = \sigma A_1$ and $m_2 = \sigma A_2$, where σ is the constant mass per unit area of the uniform sheet. Since the mass of each portion may be considered to be concentrated at its center of gravity, the coordinates of the center of gravity of the irregular plywood sheet are

$$x_{cg} = \frac{\Sigma m_i x_i}{\Sigma m_i} = \frac{(\sigma A_1)x_1 + (\sigma A_2)x_2}{\sigma A_1 + \sigma A_2} = \frac{(16.0 \text{ ft}^2)(2.00 \text{ ft}) + (8.00 \text{ ft}^2)(6.00 \text{ ft})}{16.0 \text{ ft}^2 + 8.00 \text{ ft}^2} = 3.33 \text{ ft} \qquad \lozenge$$

and

$$y_{cg} = \frac{\Sigma m_i y_i}{\Sigma m_i} = \frac{(\sigma A_1)y_1 + (\sigma A_2)y_2}{\sigma A_1 + \sigma A_2} = \frac{(16.0 \text{ ft}^2)(2.00 \text{ ft}) + (8.00 \text{ ft}^2)(1.00 \text{ ft})}{16.0 \text{ ft}^2 + 8.00 \text{ ft}^2} = 1.67 \text{ ft} \qquad \lozenge$$

19. A 500-N uniform rectangular sign 4.00 m wide and 3.00 m high is suspended from a horizontal, 6.00-m-long, uniform, 100-N rod as indicated in Figure P8.19. The left end the rod is supported by a hinge, and the right end is supported by a thin cable making a 30.0° angle with the vertical. (a) Find the tension T in the cable. (b) Find the horizontal and vertical components of force exerted on the left end of the rod by the hinge.

Solution

The free-body diagram of the sign-rod combination is given below. Note that the tension in the cable and the reaction force exerted on the left end of the rod by the hinge have been resolved into horizontal and vertical components.

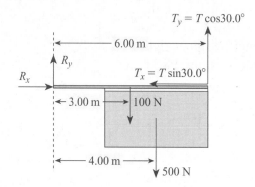

Consider a rotation axis perpendicular to the page and passing through the left end of the rod and apply the second condition of equilibrium, $\Sigma\tau = 0$, to this system. This gives

$$+(T\cos 30.0°)(6.00 \text{ m}) - (100 \text{ N})(3.00 \text{ m}) - (500 \text{ N})(4.00 \text{ m}) = 0$$

or

$$T = \frac{2.30 \times 10^3 \text{ N}\cdot\text{m}}{(6.00 \text{ m})\cos 30.0°} = 443 \text{ N} \qquad \Diamond$$

Now, apply the first condition of equilibrium to this system:

$$\Sigma F_x = 0 \ \Rightarrow\ R_x - T\sin 30.0° = 0$$

or

$$R_x = +(443 \text{ N})\sin 30.0° = +222 \text{ N} = 222 \text{ N toward the right} \qquad \Diamond$$

and

$$\Sigma F_y = 0 \ \Rightarrow\ R_y + T\cos 30.0° - 100 \text{ N} - 500 \text{ N} = 0$$

or

$$R_y = 600 \text{ N} - (443 \text{ N})\cos 30.0° = +216 \text{ N} = 216 \text{ N upward} \qquad \Diamond$$

29. The large quadriceps muscle in the upper leg terminates at its lower end in a tendon attached to the upper end of the tibia (Fig. P8.29a). The forces on the lower leg when the leg is extended are modeled as in Figure P8.29b, where $\vec{T}$ is the force of tension in the tendon, $\vec{w}$ is the force of gravity acting on the lower leg, and $\vec{F}$ is the force of gravity acting on the foot. Find $\vec{T}$ when the tendon is at an angle of 25.0° with the tibia, assuming that $w = 30.0$ N, $F = 12.5$ N, and the leg is extended at an angle θ of 40.0° with the vertical. Assume that the center of gravity of the lower leg is at its center and that the tendon attaches to the lower leg at a point one-fifth of the way down the leg.

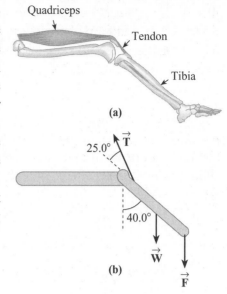

Figure P8.29

Solution

The free-body diagram of the tibia is given at the right. Note that $\theta = 40.0°$ and that all forces have been resolved into components parallel to and perpendicular to the tibia. We shall choose an axis that is perpendicular to the page and passing through the upper end of the tibia. Only T_y, w_y, and F_y have non-zero torques about this axis. The magnitudes of these components are

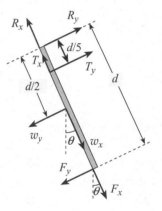

$$T_y = T\sin 25.0°$$

$$w_y = w\sin 40.0° = (30.0\ \text{N})\sin 40.0° = 19.3\ \text{N}$$

and

$$F_y = F\sin 40.0° = (12.5\ \text{N})\sin 40.0° = 8.03\ \text{N}$$

When the tibia is held in rotational equilibrium in the position shown, the second condition of equilibrium ($\Sigma\tau = 0$) gives the tension in the tendon as

$$+(T\sin 25.0°)\frac{d}{5} - (19.3\ \text{N})\frac{d}{2} - (8.03\ \text{N})d = 0$$

or

$$T = 5\left(\frac{17.7\ \text{N}}{\sin 25.0°}\right) = 209\ \text{N}$$

◊

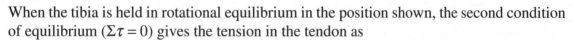

39. A 150-kg merry-go-round in the shape of a uniform, solid, horizontal disk of radius 1.50 m is set in motion by wrapping a rope about the rim of the disk and pulling on the rope. What constant force must be exerted on the rope to bring the merry-go-round from rest to an angular speed of 0.500 rev/s in 2.00 s?

Solution

The sketch at the right gives a view of the merry-go-round from above. The moment of inertia of a uniform, solid disk of mass $M = 150$ kg and radius $r = 1.50$ m is

$$I = \frac{1}{2}Mr^2 = \frac{1}{2}(150 \text{ kg})(1.5 \text{ m})^2 = 169 \text{ kg} \cdot \text{m}^2$$

The desired angular acceleration for the merry-go-round is

$$\alpha = \frac{\omega - \omega_0}{\Delta t} = \frac{0.500 \text{ rev/s} - 0}{2.00 \text{ s}} = 0.250 \frac{\text{rev}}{\text{s}^2}\left(\frac{2\pi \text{ rad}}{1 \text{ rev}}\right) = 1.57 \text{ rad/s}^2$$

We take counterclockwise torques (and hence, angular accelerations) as positive, and use the rotational form of Newton's second law to find the tension in the rope required to produce this angular acceleration:

$$\Sigma\tau = Fr = I\alpha, \quad \text{or} \quad F = \frac{I\alpha}{r} = \frac{(169 \text{ kg} \cdot \text{m})^2(1.57 \text{ rad/s}^2)}{1.50 \text{ m}} = 177 \text{ N} \qquad \lozenge$$

---

47. A solid, uniform disk of radius 0.250 m and mass 55.0 kg rolls down a ramp of length 4.50 m that makes an angle of 15.0° with the horizontal. The disk starts from rest from the top of the ramp. Find (a) the speed of the disk's center of mass when it reaches the bottom of the ramp and (b) the angular speed of the disk at the bottom of the ramp.

Solution

(a) As the disk rolls down the ramp, two non-conservative forces (a normal force and a friction force) are exerted on the disk by the ramp. The normal force is perpendicular to the motion of the disk, and hence, does no work on the disk. Assuming the disk rolls without slipping, the point on the disk in contact with the ramp at each instant is momentarily at rest, and the friction force does no work on the disk. Thus, the total mechanical energy of the disk is constant as it rolls from the top to the bottom of the ramp, or

$$\left(KE_t + KE_r + PE_g\right)_f = \left(KE_t + KE_r + PE_g\right)_i$$

Taking $y = 0$ (and hence $PE_g = 0$) at the bottom of the ramp, this becomes

$$\frac{1}{2}mv_f^2 + \frac{1}{2}I\omega_f^2 + 0 = 0 + 0 + mgy_i$$

When an object rolls without slipping, $v = r\omega$, so $\omega = v/r$. Also, $I = mr^2/2$ for a uniform, solid, disk. Our conservation of energy equation then becomes

$$\frac{1}{2}\cancel{m}v_f^2 + \frac{1}{2}\left(\frac{\cancel{m}\,\cancel{r^2}}{2}\right)\left(\frac{v^2}{\cancel{r^2}}\right) = \cancel{m}\,gy_i \qquad \text{or}$$

$$\frac{3}{4}v_f^2 = gy_i$$

and

$$v_f = \sqrt{\frac{4gy_i}{3}} = \sqrt{\frac{4(9.80\ \text{m/s}^2)(4.50\ \text{m}\cdot\sin 12.0°)}{3}} = 3.90\ \text{m/s} \qquad \diamond$$

(b) The angular speed of the disk when it reaches the bottom of the ramp is then

$$\omega_f = \frac{v_f}{r} = \frac{3.90\ \text{m/s}}{0.250\ \text{m}} = 15.6\ \text{rad/s} \qquad \diamond$$

---

53. A giant swing at an amusement park consists of a 365-kg uniform arm 10.0 m long, with two seats of negligible mass connected at the lower end of the arm (Fig. P8.53). (a) How far from the upper end is the center of mass of the arm? (b) The gravitational potential energy of the arm is the same as if all its mass were concentrated at the center of mass. If the arm is raised through a 45.0° angle, find the gravitational potential energy, where the zero level is taken to be 10.0 m below the axis. (c) The arm drops from rest. Find the gravitational potential energy of the system when it reaches the vertical orientation. (d) Find the speed of the seats at the bottom of the swing.

Solution

(a) The arm consists of a uniform rod 10.0 m long and the mass of the seats at the lower end is negligible. Thus, the center of gravity of the arm is located at the geometric center of the arm, 5.00 m from either end. $\qquad \diamond$

From the sketch at the right below, the height of the center of gravity above ground level is $y_{cg} = 10.0\ \text{m} - (5.00\ \text{m})\cos\theta$.

(b) When $\theta = 45.0°$, $y_{cg} = 10.0$ m $- (5.00$ m$)\cos 45.0° = 6.46$ m and

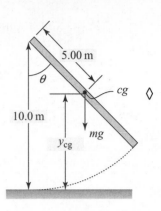

$$PE_g = mgy_{cg} = (365 \text{ kg})(9.80 \text{ m/s}^2)(6.46 \text{ m})$$
$$= 2.31 \times 10^4 \text{ J}$$

(c) In the vertical orientation, $\theta = 0°$ and $\cos\theta = 1$, giving

$$y_{cg} = 10.0 \text{ m} - 5.00 \text{ m} = 5.00 \text{ m}$$

Then,

$$PE_g = mgy_{cg} = (365 \text{ kg})(9.80 \text{ m/s}^2)(5.00 \text{ m})$$
$$= 1.79 \times 10^4 \text{ J}$$

(d) We apply conservation of mechanical energy as the arm starts from rest in the 45° orientation and rotates about an axis, perpendicular to its length and through the upper end. This gives

$$\frac{1}{2}I_{end}\omega_f^2 + mg\left(y_{cg}\right)_f = 0 + mg\left(y_{cg}\right)_i$$

or

$$\omega_f = \sqrt{\frac{2mg\left[\left(y_{cg}\right)_i - \left(y_{cg}\right)_f\right]}{I_{end}}}$$

With $I_{end} = \dfrac{mL^2}{3}$, the angular speed of the arm in the vertical orientation is

$$\omega_f = \sqrt{\frac{2mg\left[\left(y_{cg}\right)_i - \left(y_{cg}\right)_f\right]}{mL^2/3}} = \sqrt{\frac{6(9.80 \text{ m/s}^2)(6.46 \text{ m} - 5.00 \text{ m})}{(10.0 \text{ m})^2}} = 0.927 \text{ rad/s}$$

The translational speed of the seats, following a circular path of radius $r = 10.0$ m, when the arm reaches the vertical orientation is

$$v = r\omega = (10.0 \text{ m})(0.927 \text{ rad/s}) = 9.27 \text{ m/s}$$

61. A solid, horizontal cylinder of mass 10.0 kg and radius 1.00 m rotates with an angular speed of 7.00 rad/s about a fixed vertical axis through its center. A 0.250-kg piece of putty is dropped vertically onto the cylinder at a point 0.900 m from the center of rotation, and sticks to the cylinder. Determine the final angular speed of the system.

Solution

Before the putty sticks to the cylinder, the moment of inertia of this solid cylinder about an axis through its center is

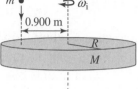

$$I_{cylinder} = \frac{1}{2}MR^2 = \frac{1}{2}(10.0 \text{ kg})(1.00 \text{ m})^2 = 5.00 \text{ kg}\cdot\text{m}^2$$

Just before its impact with the cylinder, the putty is moving parallel to the rotation axis, with no movement around that axis. Therefore, the putty has zero angular momentum about the rotation axis before impact.

The total angular momentum about the rotation axis before the putty sticks to the cylinder is then

$$L_i = L_{cylinder} + L_{putty} = I_{cylinder}\omega_i + 0 = (5.00 \text{ kg}\cdot\text{m}^2)(7.00 \text{ rad/s}) = 35.0 \text{ kg}\cdot\text{m}^2/s$$

After the putty strikes and sticks to the cylinder, both objects rotate about the rotation axis with the same angular speed ω_f. The total moment of inertia of the rotating system about the axis is now

$$I_f = I_{cylinder} + I_{putty} = 5.00 \text{ kg}\cdot\text{m}^2 + mr^2 = 5.00 \text{ kg}\cdot\text{m}^2 + (0.250 \text{ m})(0.900 \text{ m})^2$$

or

$$I_f = 5.20 \text{ kg}\cdot\text{m}^2$$

Since no external forces exert a torque about the rotation axis on the cylinder-putty system, the angular momentum of that system is conserved. Thus, $L_f = I_f\omega_f = L_i$ and

$$\omega_f = \frac{L_i}{I_f}$$

or the final angular speed of the system is

$$\omega_f = \frac{35.0 \text{ kg}\cdot\text{m}^2/s}{5.20 \text{ kg}\cdot\text{m}^2} = 6.73 \text{ rad/s} \qquad \Diamond$$

67. A 60.0-kg woman stands at the rim of a horizontal turntable having a moment of inertia of $500 \text{ kg} \cdot \text{m}^2$ and a radius of 2.00 m. The turntable is initially at rest and is free to rotate about a frictionless, vertical axle through its center. The woman then starts walking around the rim clockwise (as viewed from above the system) at a constant speed of 1.50 m/s relative to the Earth. (a) In what direction and with what angular speed does the turntable rotate? (b) How much work does the woman do to set herself and the turntable into motion?

Solution

(a) If no external agent exerts a torque about the vertical axis through the center of the turntable, the total angular momentum of the system (woman plus turntable) about this axis remains constant. When the woman is at rest on the stationary turntable, the total angular momentum of the system is zero $(L_i = 0)$. As the woman starts walking around the axis, the turntable must develop a counterclockwise angular momentum whose magnitude equals the magnitude of the woman's clockwise angular momentum. These two contributions to the total angular momentum will then cancel each other.

Treating the woman as a point object on the rim of the turntable (at distance $r = 2.00$ m from the axis), her moment of inertia about the central axis is $I_w = m_w r^2$. Taking counterclockwise angular momentum as positive, the woman's clockwise angular momentum as she walks at constant speed v relative to Earth is

$$L_w = -I_w \omega_w = -\left(m_w r^2\right)\left(\frac{v}{r}\right) = -m_w r v$$

The angular momentum of the turntable is $L_t = I_t \omega_t$ and conservation of angular momentum requires that

$$L_{\text{final}} = \left(L_t + L_w\right) = L_i = 0 \qquad \text{or} \qquad I_t \omega_t - m_w r v = 0$$

Thus,

$$\omega_t = \frac{m_w r v}{I_t} = \frac{(60.0 \text{ kg})(2.00 \text{ m})(1.50 \text{ m/s})}{500 \text{ kg} \cdot \text{m}^2} = +0.360 \text{ rad/s}$$

or

$$\omega_t = 0.360 \text{ rad/s counterclockwise} \qquad \qquad \Diamond$$

(b) The work-energy theorem gives the work done by the woman to set this system in motion as

$$W_{\text{net}} = KE_f - KE_i = \left(\tfrac{1}{2}I_t \omega_t^2 + \tfrac{1}{2}m_w v^2\right) - 0$$

$$W_{\text{net}} = \frac{1}{2}\left(500 \text{ kg} \cdot \text{m}^2\right)(0.360 \text{ rad/s})^2 + \frac{1}{2}(60.0 \text{ kg})(1.50 \text{ m/s})^2 = 99.9 \text{ J} \qquad \Diamond$$

71. A uniform ladder of length L and weight w is leaning against a vertical wall. The coefficient of static friction between the ladder and the floor is the same as that between the ladder and the wall. If this coefficient of static friction is $\mu_s = 0.500$, determine the smallest angle the ladder can make with the floor without slipping.

Solution

If the ladder is to slip, it must slip at the wall and at the floor simultaneously. Thus, at the critical angle of inclination θ, the static friction forces at the floor and wall must be at their maximum values, $(f_s)_{max} = \mu_s n$, as shown in the free-body diagram of the ladder at the right.

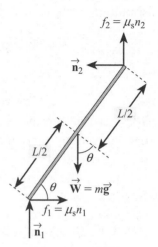

From the first condition of equilibrium,

$$\Sigma F_x = \mu_s n_1 - n_2 = 0$$

or

$$n_2 = \mu_s n_1 \qquad (1)$$

and

$$\Sigma F_y = n_1 + \mu_s n_2 - w = 0$$

or

$$n_1 + \mu_s (\mu_s n_1) = w$$

yielding

$$n_1 = \frac{w}{1 + \mu_s^2} = \frac{w}{1 + (0.500)^2}$$

or

$$n_1 = 0.800\, w \qquad (2)$$

Substitution of equation (2) into (1) gives

$$n_2 = (0.500)(0.800\, w) = 0.400\, w \qquad (3)$$

For the second condition of equilibrium, we consider the torques about an axis perpendicular to the page and passing through the lower end of the ladder. This gives

$$\Sigma \tau_{\substack{\text{lower} \\ \text{end}}} = -w\left(\frac{L}{2}\cos\theta\right) + n_2 (L\sin\theta) + \mu_s n_2 (L\cos\theta) = 0$$

Using equation (3), this becomes

$$\frac{-w}{2}\cos\theta + 0.400\, w \sin\theta + (0.500)(0.400\, w)\cos\theta = 0$$

or

$$0.400\sin\theta = (0.500 - 0.200)\cos\theta$$

and

$$\tan\theta = \frac{\sin\theta}{\cos\theta} = \frac{0.300}{0.400} = 0.750$$

The critical angle where the ladder is on the verge of slipping at the floor and at the wall simultaneously is

$$\theta = \tan^{-1}(0.750) = 36.9°$$ ◊

---

78. (a) Without the wheels, a bicycle frame has a mass of 8.44 kg. Each of the wheels can be roughly modeled as a uniform solid disk with a mass of 0.820 kg and a radius of 0.343 m. Find the kinetic energy of the whole bicycle when it is moving forward at 3.35 m/s. (b) Before the invention of a wheel turning on an axle, ancient people moved heavy loads by placing rollers under them. (Modern people use rollers, too: Any hardware store will sell you a roller bearing for a lazy Susan.) A stone block of mass 844 kg moves forward at 0.335 m/s, supported by two uniform cylindrical tree trunks, each of mass 82.0 kg and radius 0.343 m. There is no slipping between the block and the rollers or between the rollers and the ground. Find the total kinetic energy of the moving objects.

Solution

(a) The frame and the center of each wheel move forward at $v = 3.35$ m/s and each wheel also turns at angular speed $\omega = v/R$. The total kinetic energy of the bicycle is $KE = KE_t + KE_r$, or

$$KE = \frac{1}{2}(m_{frame} + 2m_{wheel})v^2 + 2\left(\frac{1}{2}I_{wheel}\omega^2\right)$$

$$= \frac{1}{2}(m_{frame} + 2m_{wheel})v^2 + \frac{1}{2}(m_{wheel}R^2)\left(\frac{v^2}{R^2}\right)$$

This yields

$$KE = \frac{1}{2}(m_{frame} + 3m_{wheel})v^2 = \frac{1}{2}[8.44 \text{ kg} + 3(0.820 \text{ kg})](3.35 \text{ m/s})^2 = 61.2 \text{ J}$$ ◊

(b) Since the block does not slip on the roller, its forward speed must equal that of point A, the uppermost point on the rim of the roller. That is, $v = |\vec{v}_{AE}|$ where $\vec{v}_{AE}$ is the velocity of A relative to the ground.

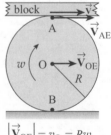

Since the roller does not slip on the ground, the velocity of point O (the roller center) relative to the ground must have the same magnitude as the tangential speed of a point on the rim of the roller. That is, $|\vec{v}_{OE}| = R\omega = v_O$. Also, note that the velocity of point A relative to the roller center has a magnitude equal to the tangential speed $R\omega$, or $|\vec{v}_{AO}| = R\omega = v_O$.

$$|\vec{v}_{OE}| = v_O = Rw$$

From the discussion of relative velocities in Chapter 3, we know that $\vec{v}_{AO} = \vec{v}_{AE} - \vec{v}_{OE}$, or the velocity of point A relative to the ground is $\vec{v}_{AE} = \vec{v}_{AO} + \vec{v}_{OE}$. Since all of these velocities are in the same direction, we may add their magnitudes getting $\left|\vec{v}_{AE}\right| = \left|\vec{v}_{AO}\right| + \left|\vec{v}_{OE}\right|$, or $v = v_O + v_O = 2v_O = 2R\omega$. Thus, we have determined that the translational speed of the center of the rollers (trees) is one half the speed of the stone $\left(v_O = v/2\right)$, and the angular speed of the roller is $\omega = v_O/R = v/2R$.

The total kinetic energy is $KE = KE_{\text{translation}} + KE_{\text{rotation}}$, or

$$KE = \frac{1}{2}m_{\text{stone}}v^2 + 2\left[\frac{1}{2}m_{\text{tree}}\left(\frac{v}{2}\right)^2\right] + 2\left(\frac{1}{2}I_{\text{tree}}\omega^2\right)$$

$$= \left(\frac{1}{2}m_{\text{stone}} + \frac{1}{4}m_{\text{tree}}\right)v^2 + \frac{1}{2}m_{\text{tree}}R^2\left(\frac{v^2}{4R^2}\right) = \frac{1}{2}\left(m_{\text{stone}} + \frac{3}{4}m_{\text{tree}}\right)v^2$$

This gives $KE = \frac{1}{2}\left[844 \text{ kg} + \frac{3}{4}\left(82.0 \text{ kg}\right)\right]\left(0.335 \text{ m/s}\right)^2 = 50.8 \text{ J}$ ◊

---

84. A string is wrapped around a uniform cylinder of mass M and radius R. The cylinder is released from rest with the string vertical and its top end tied to a fixed bar (Fig. P8.84). Show that (a) the tension in the string is one-third the weight of the cylinder, (b) the magnitude of the acceleration of the center of gravity is $2g/3$, and (c) the speed of the center of gravity is $(4gh/3)^{1/2}$ after the cylinder has descended through distance h. Verify your answer to (c) with the energy approach.

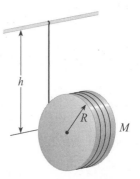

Figure P8.84

Solution

(a) The cylinder will rotate about an axis perpendicular to the page and through its center of gravity as the center of gravity accelerates downward. Applying Newton's second law to both the translational motion and the rotational motion gives:

From $\Sigma F_y = ma_y$, taking upward as positive,

$$T - Mg = M\left(-a\right)$$

or

$$T = M\left(g - a\right) \tag{1}$$

From $\Sigma\tau = I\alpha$, taking counterclockwise as positive,

$$-TR = \left(\frac{1}{2}MR^2\right)\left(-\frac{a}{R}\right)$$

or

$$a = \frac{2T}{M} \qquad\qquad (2)$$

Substituting equation (2) into (1) and solving for the tension yields

$$T = Mg/3 \qquad\qquad \Diamond$$

(b) From Equation (2),

$$a = \frac{2T}{M} = \frac{2(Mg/3)}{M} = 2g/3 \qquad\qquad \Diamond$$

(c) From $v_y^2 = v_{0y}^2 + 2a_y\Delta y$,

$$v = \sqrt{0 + 2\left(-\frac{2g}{3}\right)(-h)} = \sqrt{\frac{4gh}{3}} \qquad\qquad \Diamond$$

Using the work-energy theorem, $W_{net} = KE_f - KE_i$, we have

$$Mgh - Th = \frac{1}{2}Mv^2 - 0$$

or

$$v = \sqrt{\frac{2Mgh - 2(Mg/3)h}{M}} = \sqrt{\frac{4gh}{3}} \qquad\qquad \Diamond$$

9

Solids and Fluids

NOTES FROM SELECTED CHAPTER SECTIONS

9.1 States of Matter

Matter is generally classified as being in one of three states: solid, liquid, or gas.

In a **crystalline solid,** the atoms are arranged in an ordered periodic structure; while in an **amorphous solid** (i.e., glass), the atoms are present in a disordered fashion.

In the **liquid state,** thermal agitation is greater than in the solid state, the molecular forces are weaker, and molecules wander throughout the liquid in a random fashion.

The molecules of a **gas** are in constant random motion and exert weak forces on each other. The distances separating molecules are large compared to the dimensions of the molecules.

9.2 The Deformation of Solids

The **elastic properties of solids** are described in terms of stress and strain. Stress is a quantity that is related to the force causing a deformation; strain is a measure of the degree of deformation. It is found that, for sufficiently small stresses, stress is proportional to strain, and the constant of proportionality depends on the material being deformed and on the nature of the deformation. We call this proportionality constant the elastic modulus.

We shall consider three types of deformation and define an elastic modulus for each:

- **Young's modulus,** which measures the resistance of a solid to a change in its length

- **Shear modulus,** which measures the resistance to displacement of the planes of a solid sliding past each other

- **Bulk modulus,** which measures the resistance that solids or liquids offer to changes in their volume

9.3 Density and Pressure

The **density,** ρ, of a substance of uniform composition is defined as its mass per unit volume and has units of kilograms per cubic meter (kg/m^3) in the SI system.

The **specific gravity** of a substance is a dimensionless quantity that is the ratio of the density of the substance to the density of water.

The **pressure,** P, in a fluid is the force per unit area that the fluid exerts on an object immersed in the fluid.

9.4   Variation of Pressure with Depth

In a fluid at rest, all points at the same depth are at the same pressure. **Pascal's law** states that a change in pressure applied to an enclosed fluid is transmitted undiminished to every point in the fluid and the walls of the containing vessel. *The pressure, P, at a depth of h below the surface of a liquid with density ρ and open to the atmosphere is greater than atmospheric pressure by an amount ρgh.*

9.5   Pressure Measurements

The **absolute pressure** of a fluid is the sum of the **gauge pressure** and **atmospheric pressure.** The SI unit of pressure is the Pascal (Pa). Note that $1 \text{ Pa} \equiv 1 \text{ N/m}^2$.

9.6   Buoyant Forces and Archimedes's Principle

Any object partially or completely submerged in a fluid experiences a **buoyant force** equal in magnitude to the weight of the fluid displaced by the object and acting vertically upward through the point which was the center of gravity of the displaced fluid.

9.7   Fluids in Motion

Many features of fluid motion can be understood by considering the behavior of an ideal fluid, which satisfies the following conditions:

- **The fluid is nonviscous;** that is, there is no internal friction force between adjacent fluid layers.

- **The fluid is incompressible,** which means that its density is constant.

- **The fluid motion is steady,** meaning that the velocity, density, and pressure at each point in the fluid do not change in time.

- **The fluid moves without turbulence.** This implies that each element of the fluid has zero angular velocity about its center; that is, there can be no eddy currents present in the moving fluid.

Fluids that have the "ideal" properties stated above obey two important equations:

- The **equation of continuity** states that the flow rate through a pipe is constant (i.e., the product of the cross-sectional area of the pipe and the speed of the fluid is constant).

- **Bernoulli's equation** states that the sum of the pressure (P), kinetic energy per unit volume ($\rho v^2/2$), and the potential energy per unit volume (ρgh) has a constant value at all points along a streamline.

9.9   Surface Tension, Capillary Action, and Viscous Fluid Flow

The concept of **surface tension** can be thought of as the energy content of the fluid at its surface per unit surface area. In general, any equilibrium configuration of an object is one in which the energy is minimum. For a given volume, the spherical shape is the one that has the smallest surface area; therefore, a drop of water takes on a spherical shape. The surface tension of liquids decreases with increasing temperature.

Forces between like molecules, such as the forces between water molecules, are called **cohesive forces,** and forces between unlike molecules, such as those of glass on water, are **adhesive forces.** If a capillary tube is inserted into a fluid for which adhesive forces dominate over cohesive forces, the surrounding liquid will rise into the tube. If a capillary tube is inserted into a liquid in which cohesive forces dominate over adhesive forces, the level of the liquid in the capillary tube will be below the surface of the surrounding fluid.

Viscosity refers to the internal friction of a fluid. At sufficiently high velocities, fluid flow changes from simple streamline flow to turbulent flow. The onset of turbulence in a tube is determined by a factor called the **Reynolds number,** which is a function of the density of the fluid, the average speed of the fluid along the direction of flow, the diameter of the tube, and the viscosity of the fluid.

9.10 Transport Phenomena

The two fundamental processes involved in fluid transport resulting from concentration differences are called **diffusion** and **osmosis.** In a diffusion process, molecules move from a region where their concentration is high to a region where their concentration is lower. Diffusion occurs readily in air; the process also occurs in liquids and, to a lesser extent, in solids. Osmosis is defined as the movement of water from a region where its concentration is high, across a selectively permeable membrane, into a region where its concentration is lower. Osmosis is often described simply as the diffusion of water across a membrane.

EQUATIONS AND CONCEPTS

The elastic modulus, defined as the ratio of the *stress* to *strain,* is a general characterization of the deformation of a material. Stress is a quantity that is proportional to the force which causes the deformation, and strain is a measure of the degree of deformation. *There is an elastic modulus corresponding to each type of deformation*: *change in length, shape, and volume.*

$$\text{stress} = \text{elastic modulus} \times \text{strain} \qquad (9.1)$$

Young's modulus (Y) is a measure of the resistance of a body to elongation or compression. It is defined as the ratio of tensile stress to tensile strain. The *elastic limit* is the maximum stress from which a substance will recover to an initial length.

$$\frac{F}{A} = Y \frac{\Delta L}{L_0} \qquad (9.3)$$

The **shear modulus** is a measure of the resistance of a material to internal planes sliding past each other. Deformation occurs when a force is applied along a direction parallel to one surface of a body. As illustrated in the figure, a force $\vec{F}$ applied parallel to the top surface, with the bottom surface fixed, will cause the top surface to move forward a distance Δx.

$$\frac{F}{A} = S\frac{\Delta x}{h} \tag{9.4}$$

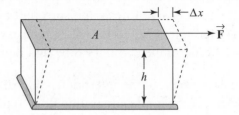

The **bulk modulus** is a measure of the resistance of a substance to uniform pressure (or squeezing) on all sides. Note that when ΔP is positive (increase in pressure), the ratio $\Delta V/V$ will be negative (decrease in volume), and vice versa. Therefore, the negative sign in the equation ensures that B will always be positive. *Compressibility of a substance is the reciprocal of the bulk modulus.*

$$\Delta P = -B\frac{\Delta V}{V} \tag{9.5}$$

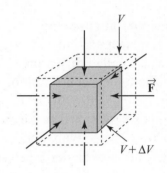

The **density** of a homogeneous substance is defined as its ratio of mass per unit volume. *Density is characteristic of a particular type of material and independent of the total quantity of material in the sample.*

$$\rho \equiv \frac{M}{V} \tag{9.6}$$

The **SI units of density** are kg per cubic meter.

$$1\ g/cm^3 = 1\,000\ kg/m^3$$

Pressure is defined as the normal force per unit area acting on a surface. Pressure has units of pascals (Pa).

$$P \equiv \frac{F}{A} \tag{9.7}$$

$$1\ Pa = 1\ N/m^2$$

Other pressure units include atmospheres, Torr, and pounds per sq. inch.

$$1\ atm = 1.013 \times 10^5\ Pa$$

$$1\ Torr = 133.3\ Pa$$

$$1\ lb/in^2 = 6.895 \times 10^3\ Pa$$

The **absolute pressure,** P, at a depth, h, below the surface of a liquid that is open to the atmosphere is greater than atmospheric pressure, P_0, by an amount which depends on the depth below the surface. *The pressure at a given depth below the surface of a liquid has the same value at all points and does not depend on the shape of the container.*

$$P = P_0 + \rho g h \qquad (9.11)$$

$$P_0 = 1.013 \times 10^5 \text{ Pa}$$

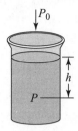

The quantity $\rho g h$ is called the *gauge pressure* and P is the *absolute pressure.*

$$\frac{\text{absolute}}{\text{pressure}} = \frac{\text{atmospheric}}{\text{pressure}} + \frac{\text{gauge}}{\text{pressure}}$$

Pascal's law states that pressure applied to an enclosed fluid (liquid or gas) is transmitted undiminished to every point within the fluid and over the walls of the vessel which contain the fluid.

Archimedes' principle states that when an object is partially or fully submerged in a fluid, the fluid exerts an upward *buoyant force, B*, which depends on the fluid density and volume of the displaced fluid. *The buoyant force equals the weight of the displaced fluid.*

$$B = \rho_{\text{fluid}} V_{\text{fluid}} g \qquad (9.12b)$$

V_{fluid} = volume of <u>displaced</u> fluid

The **ideal fluid model** is based on the following characteristics:

- **nonviscous**—internal friction between adjacent fluid layers is negligible.

- **steady flow**—the velocity, density, and pressure at each point in the fluid are constant in time.

- **incompressible**—the density throughout the fluid is constant.

- **irrotational** (without turbulence)—there are no eddy currents within the fluid (each element of the fluid has zero angular velocity about its center).

The **equation of continuity** states that *the flow rate is constant at every point along a pipe carrying an incompressible fluid.*

$$A_1 v_1 = A_2 v_2 \qquad (9.15)$$

Av = flow rate

Bernoulli's equation states that the sum of pressure, kinetic energy per unit volume, and potential energy per unit volume remains constant along a streamline of an ideal fluid. *It is a statement of the law of conservation of mechanical energy as applied to a fluid.*

$$P + \tfrac{1}{2}\rho v^2 + \rho gy = \text{constant} \qquad (9.17)$$

The **surface tension,** γ, in a film of liquid is defined as the ratio of the magnitude of the surface tension force, F, to the length, L, along which the force acts.

$$\gamma \equiv \frac{F}{L} \qquad (9.19)$$

In a **capillary tube,** liquid will rise (when adhesive forces are greater than cohesive forces) or be depressed (when cohesive forces are greater) relative to the surface of the surrounding liquid. The contact angle (ϕ) is the angle between the solid surface and the tangent to the liquid at the surface.

$$h = \frac{2\gamma}{\rho gr}\cos\phi \qquad (9.22)$$

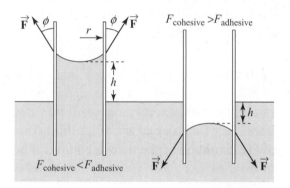

The coefficient of viscosity, η (the lowercase Greek letter eta), may be thought of as the ratio of the shearing stress to the *rate of change* of the shear strain. The SI unit of viscosity is the poise. Consider a layer of liquid of thickness d and area A between two solid surfaces. In Equation (9.23), F is the force required to move one of the solid surfaces with a velocity v relative to the other.

$$F = \eta\frac{Av}{d} \qquad (9.23)$$

$$1 \text{ poise} = 10^{-1} \text{ N}\cdot\text{s/m}^2 \qquad (9.24)$$

The **Reynolds number** is a dimensionless factor which is an indicator of the flow characteristics of a liquid in a tube. It is a function of the density of the fluid, the average speed of the fluid along the direction of flow, the diameter of the tube, and the viscosity of the fluid.

$$RN = \frac{\rho vd}{\eta} \qquad (9.26)$$

Streamline Flow:	$RN < {\sim}2\,000$
Unstable Flow:	$2\,000 < RN < 3\,000$
Turbulent Flow:	$RN > 3\,000$

The **diffusion rate** is a measure of the mass being transported per unit time. Equation (9.27) is called Fick's law.

$$\text{Diffusion Rate} = \frac{\Delta M}{\Delta t} = DA\left(\frac{C_2 - C_1}{L}\right)$$

(9.27)

D = diffusion coefficient

$(C_2 - C_1)/L$ = concentration gradient

Stokes' law describes the resistive force on a small spherical object of radius r falling with speed v through a viscous fluid.

$$F_r = 6\pi\eta r v$$

(9.28)

Terminal speed is achieved as a sphere of radius r falls through a viscous medium. The sphere is acted on by three forces: the force of frictional resistance, the buoyant force of the fluid, and the weight of the sphere. When the net upward force balances the downward weight force, the sphere reaches terminal speed.

$$v_t = \frac{2r^2 g}{9\eta}\left(\rho - \rho_f\right)$$

(9.29)

v_t = terminal speed.

In a **centrifuge** those particles having the greatest mass will have the largest terminal speed. Therefore, the most massive particles will settle out of the mixture. The factor k is a coefficient of frictional resistance which must be determined experimentally.

$$v_t = \frac{m\omega^2 r}{k}\left(1 - \frac{\rho_f}{\rho}\right)$$

(9.32)

REVIEW CHECKLIST

- Describe the three types of deformations that can occur in a solid, and define the elastic modulus that is used to characterize each (Young's modulus, shear modulus, and bulk modulus).

- Understand the concept of pressure at a point in a fluid, and the variation of pressure with depth. Understand the relationships among absolute, gauge, and atmospheric pressure values; and know the several different units commonly used to express pressure.

- Understand the origin of buoyant forces; and state and explain Archimedes' principle.

- State and understand the physical significance of the equation of continuity (constant flow rate) and Bernoulli's equation for fluid flow (relating flow velocity, pressure, and pipe elevation).

SOLUTIONS TO SELECTED END-OF-CHAPTER PROBLEMS

4. When water freezes, it expands about 9.00%. What would be the pressure increase inside your automobile engine block if the water in it froze? The bulk modulus of ice is 2.00×10^9 N/m^2.

Solution

We assume that the coolant space inside the engine block was completely filled with water. In the liquid state, this water occupied some volume, V_{liquid}. When frozen, the water would normally occupy a volume $1.090\,0\,V_{liquid}$. However, unless the engine block bursts, sufficient pressure builds up to compress the ice and force it to occupy volume V_{liquid}.

Consider the increase in pressure required to compress ice into the space it would occupy in the liquid state. Then, the original volume of the ice is $V_0 = 1.090\,0\,V_{liquid}$, the final volume is $V_f = V_{liquid}$, and the change in volume is

$$\Delta V = V_f - V_0 = V_{liquid} - 1.090\,0\,V_{liquid} = -0.090\,0\,V_{liquid}$$

From the definition of bulk modulus,

$$B = -\frac{\Delta P}{\Delta V / V_0}$$

The increase in pressure required to accomplish this compression is

$$\Delta P = -B\left(\frac{\Delta V}{V_0}\right) = -\left(2.00 \times 10^9 \text{ Pa}\right)\left(\frac{-0.090\,0\,V_{liquid}}{1.090\,0\,V_{liquid}}\right) = 1.65 \times 10^8 \text{ Pa} = 165 \text{ MPa} \qquad \Diamond$$

Note that this increase in pressure is

$$\Delta P = \left(1.65 \times 10^8 \text{ Pa}\right)\left(\frac{1 \text{ atm}}{1.013 \times 10^5 \text{ Pa}}\right) \approx 1\,600 \text{ atm}$$

Thus, there is a good possibility that the engine block will burst before the compression is accomplished.

11. Determine the elongation of the rod in Figure P9.11 if it is under a tension of 5.8×10^3 N.

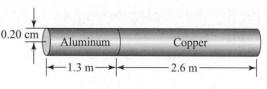

Figure P9.11

Solution

When the rod comes to equilibrium, the tension will be uniform throughout its length, with both types of metal subject to a stretching force of $F = 5.8 \times 10^3$ N. Also, the cross-sectional area is uniform throughout the rod, with a value of

$$A = \pi r^2 = \pi \left(0.20 \times 10^{-2} \text{ m}\right)^2 = \pi \left(2.0 \times 10^{-3} \text{ m}\right)^2 = 4.0\pi \times 10^{-6} \text{ m}^2$$

The total elongation of the rod is the sum of the elongation of the aluminum section and that of the copper section, or $\Delta L_{rod} = \Delta L_{Al} + \Delta L_C$.

Young's modulus of a material is given by

$$Y = \frac{tensil\ stress}{tensil\ strain} = \frac{F/A}{\Delta L/L_0}$$

Here, F is the force of tension or compression acting on the material, A is the cross-sectional area, L_0 is the original length of the material, and ΔL is the elongation the material undergoes. Hence, the elongation of each section of this rod is given by $\Delta L = FL_0/YA$, and the total elongation of the rod becomes

$$\Delta L_{rod} = \frac{F\left(L_0\right)_{Al}}{Y_{Al}A} + \frac{F\left(L_0\right)_{Cu}}{Y_{Cu}A} = \frac{F}{A}\left[\frac{\left(L_0\right)_{Al}}{Y_{Al}} + \frac{\left(L_0\right)_{Cu}}{Y_{Cu}}\right]$$

or

$$\Delta L_{rod} = \frac{5.8 \times 10^3 \text{ N}}{4.0\pi \times 10^{-6} \text{ m}^2}\left[\frac{1.3 \text{ m}}{7.0 \times 10^{10} \text{ Pa}} + \frac{2.6 \text{ m}}{11 \times 10^{10} \text{ Pa}}\right] = 1.9 \times 10^{-2} \text{ m} = 1.9 \text{ cm} \qquad \Diamond$$

19. If 1.0 m³ of concrete weighs 5.0×10^4 N, what is the height of the tallest cylindrical concrete pillar that will not collapse under its own weight? The compression strength of concrete (the maximum pressure that can be exerted on the base of the structure) is 1.7×10^7 Pa.

Solution

Note that the product $g\rho_{concrete}$, where $\rho_{concrete}$ is the mass per unit volume (density) for concrete, gives the weight per unit volume. We assume the cylindrical pillar has height h and cross-sectional area A. Hence, the volume is $V = Ah$, and the weight of the pillar will be

$$F_g = \left(g\rho_{concrete}\right)V = \left(g\rho_{concrete}\right)Ah$$

The pressure on the concrete forming the base of the structure is then

$$P = \frac{F_g}{A} = \frac{\left(g\rho_{concrete}\right)Ah}{A} = \left(g\rho_{concrete}\right)h$$

If the maximum pressure concrete can withstand is $P_{max} = 1.7 \times 10^7$ Pa, the maximum height the pillar can have without collapsing under its own weight is

$$h_{max} = \frac{P_{max}}{\left(g\rho_{concrete}\right)} = \frac{1.7 \times 10^7 \text{ Pa}}{5.0 \times 10^4 \text{ N/m}^3} = 3.4 \times 10^2 \text{ m} \approx 1\ 100 \text{ ft}$$ ◊

23. A collapsible plastic bag (Figure P9.23) contains a glucose solution. If the average gauge pressure in the vein is 1.33×10^3 Pa, what must be the minimum height h of the bag in order to infuse glucose into the vein? Assume the specific gravity of the solution is 1.02.

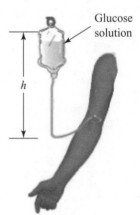

Figure P9.23

Solution

If the glucose solution has a specific gravity of 1.02, its density is

$$\rho_{gs} = 1.02 \cdot \rho_{water} = 1.02\left(1.00 \times 10^3 \text{ kg/m}^3\right) = 1.02 \times 10^3 \text{ kg/m}^3$$

The gauge pressure in the vein is $\left(P_{gauge}\right)_{vein} = 1.33 \times 10^3$ Pa, so in order to infuse glucose into the vein, the gauge pressure in the glucose solution at the point where it is to enter the vein must be

$$\left(P_{gauge}\right)_{gs} \geq \left(P_{gauge}\right)_{vein} = 1.33 \times 10^3 \text{ Pa}$$

Since the glucose solution container is a collapsible plastic bag, the bag will collapse and maintain atmospheric pressure at the upper surface of the solution as the solution drains out of the bag. The pressure at depth h below this upper surface (i.e., at the level where the glucose enters the vein) is given by $P = P_{atm} + \rho_{gs}gh$, and the gauge pressure the entry point to the vein is $\left(P_{gauge}\right)_{gs} = P - P_{atm} = \rho_{gs}gh$. Thus, for infusion into the vein, it is necessary to have $\rho_{gs}gh \geq 1.33 \times 10^3$ Pa, or

$$h \geq \frac{1.33 \times 10^3 \text{ Pa}}{\rho_{gs}g} = \frac{1.33 \times 10^3 \text{ Pa}}{\left(1.02 \times 10^3 \text{ kg/m}^3\right)\left(9.80 \text{ m/s}^2\right)} = 0.133 \text{ m}$$ ◊

27. Figure P9.27 shows the essential parts of a hydraulic brake system. The area of the piston in the master cylinder is 1.8 cm^2 and that of the piston in the brake cylinder is 6.4 cm^2. The coefficient of friction between shoe and wheel drum is 0.50. If the wheel has a radius of 34 cm, determine the frictional torque about the axle when a force of 44 N is exerted on the brake pedal.

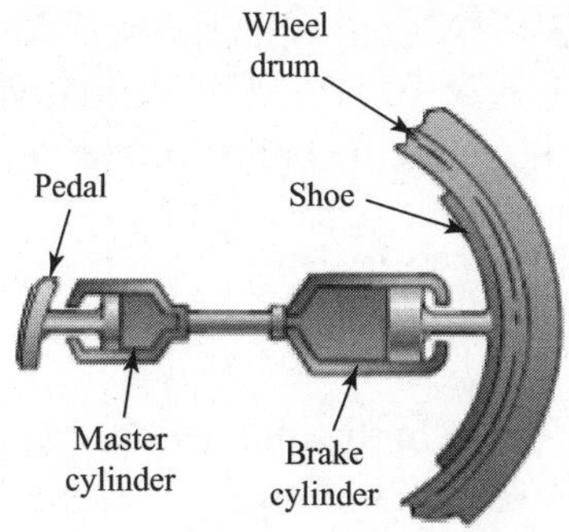

Figure P9.27

Solution

When the brake pedal exerts a force of $F_1 = 44 \text{ N}$ on the piston in the master cylinder, it creates an increase in pressure of

$$\Delta P = \frac{F_1}{A_1} = \frac{44 \text{ N}}{1.8 \text{ cm}^2}\left(\frac{10^4 \text{ cm}^2}{1 \text{ m}^2}\right) = 2.4 \times 10^5 \text{ Pa}$$

in the enclosed fluid of the hydraulic brake system. According to Pascal's principle, this change in pressure is transmitted undiminished to every point of the fluid and to the walls of the container. Thus, an increase in pressure of $\Delta P = 2.4 \times 10^5$ Pa also occurs in the fluid inside the brake cylinder and this fluid will exert an increased force of

$$F_2 = (\Delta P)A_2 = (2.4 \times 10^5 \text{ Pa})(6.4 \text{ cm}^2)\left(\frac{1 \text{ m}^2}{10^4 \text{ cm}^2}\right) = 1.6 \times 10^2 \text{ N}$$

to the piston of the brake cylinder, and hence as a normal force to the brake shoe. The friction force between the brake shoe and the wheel drum will then be

$$f_k = \mu_k n = (0.50)(1.6 \times 10^2 \text{ N}) = 80 \text{ N}$$

This friction force between the shoe and drum will produce a frictional torque about the axle of

$$\tau = f_k \cdot r = (80 \text{ N})(0.34 \text{ m}) = 27 \text{ N} \cdot \text{m}$$

◊

39. A bathysphere used for deep sea exploration has a radius of 1.50 m and a mass of 1.20×10^4 kg. In order to dive, the sphere takes on mass in the form of sea water. Determine the mass the bathysphere must take on so that it can descend at a constant speed of 1.20 m/s when the resistive force on it is 1 100 N upward. The density of sea water is 1.03×10^3 kg/m³.

Solution

As the bathysphere descends, there are four forces acting on it as shown in the diagram at the right. These forces are: (1) the upward buoyant force $\vec{\mathbf{B}}$ exerted on the bathysphere by the water; (2) an upward resistance force $\vec{\mathbf{R}}$ opposing the motion of the bathysphere through the water; the weight $\vec{\mathbf{F}}_g$ of the bathysphere itself; and the weight $\left(\vec{\mathbf{F}}_g\right)_b$ of the sea water taken on board as ballast.

Bathysphere

When the submarine descends at constant speed, it is in equilibrium and $\Sigma F_y = 0$ gives

$$B + R - F_g - \left(F_g\right)_b = 0 \qquad \text{or} \qquad \left(F_g\right)_b = B + R - F_g$$

The buoyant force is

$$B = \left(\rho_{\substack{sea \\ water}} \; V_{\substack{water \\ displaced}}\right)g = \left(\rho_{\substack{sea \\ water}} \; V_{\substack{bathysphere}}\right)g = \rho_{\substack{sea \\ water}} \left(\frac{4\pi r^3}{3}\right)g$$

Thus,

$$\left(F_g\right)_b = m_{\text{ballast}}g = \rho_{\substack{sea \\ water}} \left(\frac{4\pi r^3}{3}\right)g + R - \left(m_{\text{bathysphere}}\right)g$$

and the mass of sea water that should be taken on as ballast is

$$m_{\text{ballast}} = \rho_{\substack{sea \\ water}} \left(\frac{4\pi r^3}{3}\right) + \frac{R}{g} - m_{\text{bathysphere}}$$

or

$$m_{\text{ballast}} = \left(1.03 \times 10^3 \; \text{kg/m}^2\right)\frac{4\pi (1.50 \; \text{m})^3}{3} + \left(\frac{1\,100 \; \text{N}}{9.80 \; \text{m/s}^2}\right) - 1.20 \times 10^4 \; \text{kg}$$

yielding

$$m_{\text{ballast}} = 2.67 \times 10^3 \; \text{kg}$$

◊

43. A 1.00-kg beaker containing 2.00 kg of oil (density = 916 kg/m^3) rests on a scale. A 2.00-kg block of iron is suspended from a spring scale and is completely submerged in the oil (Fig. P9.43). Find the equilibrium readings of both scales.

Figure P9.43

Solution

The volume of the iron block is

$$V = \frac{m_{iron}}{\rho_{iron}} = \frac{2.00 \text{ kg}}{7.86 \times 10^3 \text{ kg/m}^3} = 2.54 \times 10^{-4} \text{ m}^3$$

and the buoyant force exerted on the iron by the oil is

$$B = (\rho_{oil} V)g = (916 \text{ kg/m}^3)(2.54 \times 10^{-4} \text{ m}^3)(9.80 \text{ m/s}^2) = 2.28 \text{ N}$$

Applying $\Sigma F_y = 0$ to the iron block gives the support force exerted by the upper spring scale (and hence the reading on that scale) as

$$F_{upper} = m_{iron} g - B = (2.00 \text{ kg})(9.80 \text{ m/s}^2) - 2.28 \text{ N} = 17.3 \text{ N} \qquad \Diamond$$

From Newton's third law, the iron exerts a reaction force of magnitude B downward on the oil (and hence the beaker). Other vertical forces acting on the beaker are (1) the combined weight of the beaker and oil, and (2) an upward support force exerted by the lower scale. Applying $\Sigma F_y = 0$ to the system consisting of the beaker and the oil gives

$$F_{lower} - B - (m_{oil} + m_{beaker})g = 0$$

The support force exerted by the lower scale (and the lower scale reading) is then

$$F_{lower} = B + (m_{oil} + m_{beaker})g = 2.28 \text{ N} + [(2.00 + 1.00) \text{ kg}](9.80 \text{ m/s}^2) = 31.7 \text{ N} \qquad \Diamond$$

47. A hypodermic syringe contains a medicine with the density of water (Fig. P9.47). The barrel of the syringe has a cross-sectional area of 2.50×10^{-5} m^2. In the absence of a force on the plunger, the pressure everywhere is 1.00 atm. A force $\vec{F}$ of magnitude 2.00 N is exerted on the plunger, making medicine squirt from the needle. Determine the medicine's flow speed through the needle. Assume the pressure in the needle remains equal to 1.00 atm and that the syringe is horizontal.

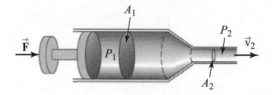

Figure P9.47

Solution

The pressure inside the needle, which is open to the atmosphere, remains at $P_2 = 1.00$ atm until force is applied to the plunger of the needle. Applying a force of $F = 2.00$ N to the plunger produces an increase in pressure inside the syringe chamber of

$$\Delta P = P_1 - 1.00 \text{ atm} = P_1 - P_2 = \frac{F}{A_1} = \frac{2.00 \text{ N}}{2.50 \times 10^{-5} \text{ m}^2} = 8.00 \times 10^4 \text{ Pa}$$

From Bernoulli's equation, $P_1 + \frac{1}{2}\rho_w v_1^2 + \rho_w g y_1 = P_2 + \frac{1}{2}\rho_w v_2^2 + \rho_w g y_2$, with $y_1 = y_2$ since the syringe is horizontal, and $v_1 \approx 0$ since the diameter of the syringe chamber is much larger than that of the opening to the needle, we have

$$v_2^2 = v_1^2 + \frac{2(P_1 - P_2)}{\rho_w} + 2g(y_1 - y_2) \approx 0 + \frac{2(P_1 - P_2)}{\rho_w} + 0$$

or the medicine's flow speed through the needle is

$$v_2 = \sqrt{\frac{2(P_1 - P_2)}{\rho_w}} = \sqrt{\frac{2(8.00 \times 10^4 \text{ Pa})}{1.00 \times 10^3 \text{ kg/m}^3}} = 12.6 \text{ m/s}$$

◊

57. Old Faithful Geyser in Yellowstone Park erupts at approximately 1-hour intervals, and the height of the fountain reaches 40.0 m. (a) Consider the rising stream as a series of separate drops. Analyze the free-fall motion of one of the drops to determine the speed at which the water leaves the ground. (b) Treat the rising stream as an ideal fluid in stream-line flow. Use Bernoulli's equation to determine the speed of the water as it leaves ground level. (c) What is the pressure (above atmospheric pressure) in the heated underground chamber 175 m below the vent? You may assume the chamber is large compared with the geyser vent.

Solution

(a) Consider the motion of a water-drop projectile as it goes from the geyser vent to the top of the fountain. We use $v_y^2 = v_{0y}^2 + 2a_y \Delta y$, with $v_y = 0$ when Δy equals the height h of the fountain. Then, the speed of the drop as it emerges from the vent is

$$v_{vent} = v_{0y} = \sqrt{v_y^2 - 2(-g)h} = \sqrt{0 - 2(-9.80 \text{ m/s}^2)(40.0 \text{ m})} = 28.0 \text{ m/s} \qquad \Diamond$$

(b) Because of the low density of air and the small change in altitude, we neglect any change in atmospheric pressure in going from the ground to the top of the fountain. Then, applying Bernoulli's equation from the vent to the top of the fountain gives

$$P_{vent} + \tfrac{1}{2}\rho_{water} v_{vent}^2 + \rho_{water} g y_{vent} = P_{top} + \tfrac{1}{2}\rho_{water} v_{top}^2 + \rho_{water} g y_{top}$$

Thus,

$$v_{vent} = \sqrt{v_{top}^2 + \frac{2}{\rho_{water}}\left[\left(P_{top} - P_{vent}\right) + \rho_{water} g\left(y_{top} - y_{vent}\right)\right]}$$

with

$$P_{top} \approx P_{vent}, \ v_{top} = 0, \text{ and } \left(y_{top} - y_{vent}\right) = h$$

or

$$v_{vent} = \sqrt{2gh} = \sqrt{2(9.80 \text{ m/s}^2)(40.0 \text{ m})} = 28.0 \text{ m/s} \qquad \Diamond$$

(c) If the chamber may be considered large in comparison to the geyser vent, we may assume that the speed of the water in the chamber is negligible in comparison to its speed at the geyser vent. Then, applying Bernoulli's equation between the chamber and geyser vent gives

$$P_{chamber} + 0 + \rho_{water} g y_{chamber} = P_{vent} + \tfrac{1}{2}\rho_{water} v_{vent}^2 + \rho_{water} g y_{vent}$$

With $P_{vent} = P_{atmo}$, this gives the gauge pressure in the chamber as

$$P_{gauge} = \left(P_{chamber} - P_{atmo}\right) = \rho_{water}\left[\tfrac{1}{2} v_{vent}^2 + g\left(y_{vent} - y_{chamber}\right)\right]$$

or

$$P_{gauge} = \left(1.00 \times 10^3 \text{ kg/m}^3\right)\left[\tfrac{1}{2}(28.0 \text{ m/s})^2 + (9.80 \text{ m/s}^2)(175 \text{ m})\right]$$

This yields

$$P_{gauge} = 2.11 \times 10^6 \text{ Pa} = 2.11 \text{ MPa} = 20.8 \text{ atmospheres} \qquad \Diamond$$

67. Spherical particles of a protein of density 1.8 g/cm³ are shaken up in a solution of 20°C water. The solution is allowed to stand for 1.0 h. If the depth of water in the tube is 5.0 cm, find the radius of the largest particles that remain in solution at the end of the hour.

Solution

Note that the density of the protein is

$$\rho = 1.8 \ \frac{g}{cm^3}\left(\frac{1 \text{ kg}}{10^3 \text{ g}}\right)\left(\frac{10^2 \text{ cm}}{1 \text{ m}}\right)^3 = 1.8 \times 10^3 \ kg/m^3$$

If a particle is still in suspension after one hour, then its terminal speed must be

$$v_t \le \left(\frac{5.0 \text{ cm}}{1.0 \text{ h}}\right)\left(\frac{1 \text{ m}}{100 \text{ cm}}\right)\left(\frac{1 \text{ h}}{3\,600 \text{ s}}\right) = 1.4 \times 10^{-5} \ m/s$$

In general, the terminal speed of a particle of density ρ and radius r falling through a fluid of density ρ_f and viscosity η is given by Equation 9.29 in the textbook as

$$v_t = \frac{2r^2 g}{9\eta}\left(\rho - \rho_f\right)$$

Thus, if the upper limit of the terminal velocity of the particles still suspended in the 20°C water is $v_{t,max} = 1.4 \times 10^{-5}$ m/s, the maximum radius of these particles is

$$r_{max} = \sqrt{\frac{9\eta \, v_{t,max}}{2g\left(\rho - \rho_f\right)}} = \sqrt{\frac{9\left(1.0 \times 10^{-3} \ N\cdot s/m^2\right)\left(1.4 \times 10^{-5} \ m/s\right)}{2\left(9.8 \ m/s^2\right)\left[\left(1.8 - 1.0\right) \times 10^3 \ kg/m^3\right]}}$$

or

$$r_{max} = 2.8 \times 10^{-6} \ m = 2.8 \ \mu m \qquad \qquad \Diamond$$

77. An iron block of volume 0.20 m^3 is suspended from a spring scale and immersed in a flask of water. Then the iron block is removed, and an aluminum block of the same volume replaces it. (a) In which case is the buoyant force the greatest, for the iron block or the aluminum block? (b) In which case does the spring scale read the largest value? (c) Use the known densities of these materials to calculate the quantities requested in parts (a) and (b). Are your calculations consistent with your previous answers to parts (a) and (b)?

Solution

(a) The buoyant force acting on an object partially or wholly submerged in a fluid of density ρ_f is equal to the weight of the displaced fluid, or $B = \rho_f g V_{\text{displaced}}$. Both iron and aluminum are denser than water, so both blocks will be fully submerged. Since the two blocks have the same volume, they displace equal volumes of water, and will experience buoyant forces of equal magnitudes. ◊

(b) The force exerted on the block by the spring scale (of magnitude equal to the scale reading) and the buoyant force act together to support the weight of the block (i.e., $F_s + B = w_{\text{block}}$ giving $F_s = w_{\text{block}} - B$). Since the volumes and the buoyant forces are the same for the two blocks, the higher spring scale reading occurs for the block having the higher density, and hence greater weight. This is the iron block, with $\rho_{\text{Fe}} > \rho_{\text{Al}}$. ◊

(c) The buoyant force in each case is

$$B = \rho_{\text{water}} g V_{\text{block}} = \left(1.00 \times 10^3 \ \text{kg/m}^3\right)\left(9.80 \ \text{m/s}^2\right)\left(0.20 \ \text{m}^3\right) = 2.0 \times 10^3 \ \text{N}$$ ◊

The scale reading for the iron block is $\left(F_s\right)_{\text{Fe}} = m_{\text{Fe}} g - B = \rho_{\text{Fe}} g V - B$, or

$$\left(F_s\right)_{\text{Fe}} = \left(7.86 \times 10^3 \ \text{kg/m}^3\right)\left(9.8 \ \text{m/s}^2\right)\left(0.20 \ \text{m}^3\right) - 2.0 \times 10^3 \ \text{N} = 1.3 \times 10^4 \ \text{N}$$ ◊

while that for the aluminum block is $\left(F_s\right)_{\text{Al}} = m_{\text{Al}} g - B = \rho_{\text{Al}} g V - B$, or

$$\left(F_s\right)_{\text{Al}} = \left(2.70 \times 10^3 \ \text{kg/m}^3\right)\left(9.8 \ \text{m/s}^2\right)\left(0.20 \ \text{m}^3\right) - 2.0 \times 10^3 \ \text{N} = 3.3 \times 10^3 \ \text{N}$$ ◊

all of which is consistent with previous answers in parts (a) and (b).

81. The approximate inside diameter of the aorta is 0.50 cm; that of a capillary is 10 μm. The approximate average blood flow speed is 1.0 m/s in the aorta and 1.0 cm/s in the capillaries. If all the blood in the aorta eventually flows through the capillaries, estimate the number of capillaries in the circulatory system.

Solution

If the diameters of the aorta and of a capillary are d_1 and d_2, respectively,

their cross-sectional areas are

$$A_{aorta} = A_1 = \frac{\pi d_1^2}{4}$$

and

$$A_{capillary} = A_c = \frac{\pi d_2^2}{4}$$

Assuming the circulatory system has a total of N capillaries, the total cross-sectional area carrying blood from the aorta is $A_2 = NA_{capillary} = N\left(\pi d_2^2\right)/4$.

The equation of continuity then requires that $A_2 v_2 = A_1 v_1$, where v_1 is the blood flow speed in the aorta and v_2 is the flow speed in a capillary.

This gives

$$N\left(\frac{\pi d_2^2}{4}\right)v_2 = \left(\frac{\pi d_1^2}{4}\right)v_1$$

so the approximate number of capillaries in the circulatory system must be

$$N = \left(\frac{d_1}{d_2}\right)^2\left(\frac{v_1}{v_2}\right) = \left(\frac{0.50\times10^{-2}\ \text{m}}{10\times10^{-6}\ \text{m}}\right)^2\left(\frac{1.0\ \text{m/s}}{1.0\times10^{-2}\ \text{m/s}}\right) = 2.5\times10^7 = 25\ \text{million} \qquad \Diamond$$

86. A helium-filled balloon is tied to a 2.0-m-long, 0.050-kg string. The balloon is spherical with a radius of 0.40 m. When released, it lifts a length h of the string and then remains in equilibrium, as in Figure P9.86. Determine the value of h. When deflated, the balloon has a mass of 0.25 kg. *Hint:* Only that part of the string above the floor contributes to the load being held up by the balloon.

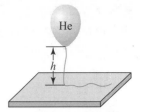

Figure P9.86

Solution

The upward buoyant force acting on the balloon is equal to the weight of air displaced by the balloon, or $B = \rho_{air} g V_{balloon}$. In addition to this force, there are three downward forces acting on the balloon. These are: (1) the weight of the elastic material making up the balloon itself, $w_{balloon} = m_{balloon} g$; (2) the weight of the helium filling the balloon, $w_{He} = m_{He} g = \rho_{He} g V_{balloon}$; and (3) the weight of the length h of string the balloon has lifted off the floor, $w_{string} = m_{lifted\ string} g = \left(\lambda_{string} h\right) g$, where λ_{string} is the mass per unit length of the string with a value of

$$\lambda_{string} = \frac{m_{total}^{length}}{L_{string}} = \frac{0.050 \text{ kg}}{2.0 \text{ m}} = 2.5 \times 10^{-2} \text{ kg/m}$$

When the balloon comes to equilibrium, $\Sigma F_y = 0$, giving $B - w_{balloon} - w_{He} - w_{string} = 0$, or

$$\rho_{air}\, \cancel{g} V_{balloon} = m_{balloon}\, \cancel{g} + \rho_{He}\, \cancel{g} V_{balloon} + \left(\lambda_{string}\, \cancel{g}\right) h$$

and the length of string lifted above the floor will be

$$h = \frac{\left(\rho_{air} - \rho_{He}\right)\left(4\pi r_{balloon}^3 / 3\right) - m_{balloon}}{\lambda_{string}}$$

and, using density values from Table 9.3 in the textbook,

$$h = \frac{\left[\left(1.29 - 0.179\right) \text{ kg/m}^3\right]\left(4\pi/3\right)\left(0.40 \text{ m}\right)^3 - 0.25 \text{ kg}}{2.5 \times 10^{-2} \text{ kg/m}} = 1.9 \text{ m}$$

◊

91. A water tank open to the atmosphere at the top has two small holes punched in its side, one above the other. The holes are 5.00 cm and 12.0 cm above the floor. How high does water stand in the tank if the two streams of water hit the floor at the same place?

Solution

A water droplet emerging from the tank becomes a projectile with initial velocity components of $v_{0y} = 0$, and $v_{0x} = v$, where v is the speed of water passing through the one of the holes in the side of the tank. The time required for this droplet to fall distance h to the floor is given by $\Delta y = v_{0y}t + a_y t^2/2$, with $\Delta y = -h$, $v_{0y} = 0$, and $a_y = -g$, as $t = \sqrt{2h/g}$. The horizontal range of the droplet is then $R = v_{0x}t = v\sqrt{2h/g}$. If the two streams are to hit the floor at the same spot, it is necessary that $R_1 = R_2$, or

$$v_1\sqrt{\frac{2h_1}{g}} = v_2\sqrt{\frac{2h_2}{g}}$$

so

$$v_1 = v_2\sqrt{\frac{h_2}{h_1}} = v_2\sqrt{\frac{12.0 \text{ cm}}{5.00 \text{ cm}}}$$

and

$$v_1^2 = 2.40 \cdot v_2^2 \qquad\qquad (1)$$

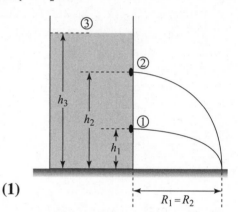

Recognize that the tank is open to the atmosphere at both points 1 (the lower hole) and 3 (the upper surface of the water in the tank), so that $P_1 = P_3 = P_{atm}$. Also, assuming the tank is very large in comparison to the size of the openings in its side, $v_3 \approx 0$. Then, writing Bernoulli's equation for this pair of points gives

$$\cancel{P_1} + \frac{1}{2}\rho_w v_1^2 + \rho_w g h_1 = \cancel{P_3} + \frac{1}{2}\rho_w (0) + \rho_w g h_3$$

or

$$v_1^2 = 2g\left(h_3 - h_1\right) \qquad\qquad (2)$$

Next, apply Bernoulli's equation to points 2 (the upper hole) and 3 (water surface), realizing the $P_2 = P_3 = P_{atm}$ and $v_3 \approx 0$, to obtain

$$\cancel{P_2} + \frac{1}{2}\rho_w v_2^2 + \rho_w g h_2 = \cancel{P_3} + \frac{1}{2}\rho_w (0) + \rho_w g h_3$$

or

$$v_2^2 = 2g\left(h_3 - h_2\right) \qquad\qquad (3)$$

Substituting equations (2) and (3) into (1) gives $h_3 - h_1 = 2.40\left(h_3 - h_2\right)$, and solving for h_3 (the depth of the water inside the tank) yields

$$h_3 = \frac{2.40h_2 - h_1}{1.40} = \frac{2.40(12.0 \text{ cm}) - 5.00 \text{ cm}}{1.40} = 17.0 \text{ cm} \qquad\qquad \Diamond$$

10

Thermal Physics

NOTES FROM SELECTED CHAPTER SECTIONS

10.1 Temperature and the Zeroth Law of Thermodynamics

The zeroth law of thermodynamics (or the equilibrium law) can be stated as follows:

If bodies A and B are separately in thermal equilibrium with a third body, C, then A and B will be in thermal equilibrium with each other if placed in thermal contact.

Two objects in thermal equilibrium with each other are at the same temperature.

10.2 Thermometers and Temperature Scales

The physical property used in a constant volume gas thermometer is the pressure variation with temperature of a fixed volume of gas. The temperature readings are nearly independent of the substance used in the thermometer.

The **triple point of water,** which is the single temperature and pressure at which water, water vapor, and ice can coexist in equilibrium, was chosen as a convenient and reproducible reference temperature for the Kelvin scale. It occurs at a temperature of $0.01°C$ and a pressure of 4.58 mm of mercury. The temperature at the triple point of water on the Kelvin scale has been assigned a value of 273.16 kelvins (K). Thus, the SI unit of temperature, the kelvin, is defined as 1/273.16 of the temperature of the triple point of water.

10.3 Thermal Expansion of Solids and Liquids

Liquids generally increase in volume with increasing temperature and have volume expansion coefficients about ten times greater than those of solids. Water is an exception to this rule; as the temperature increases from $0°C$ to $4°C$, water contracts and thus its density increases. Above $4°C$, water expands with increasing temperature. *The density of water reaches its maximum value (1 000 kg/m^3) 4 degrees above the freezing point.*

10.4 Macroscopic Description of an Ideal Gas

In an **ideal gas** the atoms or molecules move randomly, the individual particles do not exert long-range forces on each other, and each particle is considered to be a "point" mass. One **mole** of any substance is that quantity of material that contains a number of particles equal to the number of atoms in 12 grams of the isotope carbon-12. **One mole of any gas contains Avogadro's number of particles (N_A) and equal volumes of all gases at the same temperature and pressure contain the same number of particles.**

10.5 The Kinetic Theory of Gases

A microscopic **model of an ideal gas** is based on the following assumptions:

- **The number of molecules is large, and the average separation between them is large** compared with their dimensions. Therefore, the molecules occupy a negligible volume compared with the volume of the container.

- **The molecules obey Newton's laws of motion, but the individual molecules move in a random fashion.** By random fashion, we mean that the molecules move in all directions with equal probability and with various speeds. This distribution of velocities does not change in time, despite the collisions between molecules.

- **The molecules undergo elastic collisions with each other.** Thus, the molecules are considered to be without structure (that is, point masses), and in the collisions both kinetic energy and momentum are conserved.

- **The forces between molecules are negligible, except during a collision.** The forces between molecules are short-range, so that the only time the molecules interact with each other is during a collision.

- **The gas under consideration is a pure gas.** That is, all molecules are identical.

- **The molecules of the gas make perfectly elastic collisions with the walls of the container.** Hence, the wall will eject as many molecules as it absorbs, and the ejected molecules will have the same average kinetic energy as the absorbed molecules.

EQUATIONS AND CONCEPTS

The **Celsius temperature** T_C is related to the Kelvin temperature (sometimes called the absolute temperature) according to Equation (10.1). *The size of a degree on the Kelvin scale equals the size of a degree on the Celsius scale.*

$$T_C = T - 273.15 \tag{10.1}$$

T_C = Celsius temperature

T = Kelvin temperature

Conversion between Fahrenheit and Celsius scales is based on the freezing and boiling point temperatures of water set for each scale.

$$T_F = \tfrac{9}{5}T_C + 32 \tag{10.2a}$$

$$T_C = \tfrac{5}{9}(T_F - 32) \tag{10.2b}$$

Freezing point = 0° C = 32°F

Boiling point = 100° C = 212°F

Thermal expansion results in a change in length, surface area, and volume. In each case, the **fractional change** is proportional to the change in temperature and a thermal coefficient characteristic of the particular type of material.

α = coefficient of linear expansion

$\gamma \cong 2\alpha$ = coefficient of area expansion

$\beta \cong 3\alpha$ = coefficient of volume expansion

Length expansion (solids)

$$\Delta L = \alpha L_0 \Delta T \qquad (10.4)$$

$$\frac{\Delta L}{L_0} = \alpha(\Delta T)$$

Area expansion (solids)

$$\Delta A = \gamma A_0 \Delta T \qquad (10.5)$$

$$\frac{\Delta A}{A_0} = \gamma(\Delta T)$$

Volume expansion (solids and liquids)

$$\Delta V = \beta V_0 \Delta T \qquad (10.6)$$

$$\frac{\Delta V}{V_0} = \beta(\Delta T)$$

The **number of moles** in a sample of an element or compound is the ratio of the mass of the sample to the atomic or molar mass characteristic of that particular material. *One mole of a substance contains Avogadro's number (N_A) of particles.*

$$n = \frac{m}{\text{molar mass}} \qquad (10.7)$$

$$m_{\text{molecule}} = \frac{\text{molar mass}}{N_A}$$

$$N_A = 6.02 \times 10^{23} \; \frac{\text{particles}}{\text{mole}}$$

Boyle's law states that when a gas is maintained at constant temperature, the pressure is inversely proportional to the volume.

$$P \propto \left(\frac{1}{V}\right) \quad \text{(at constant temperature)}$$

Charles' law states that when a gas is maintained at constant pressure, volume is directly proportional to the absolute temperature.

$$V \propto T \quad \text{(at constant pressure)}$$

Gay-Lussac's law states that when the volume of a gas is held constant, the pressure is directly proportional to the absolute temperature.

$$P \propto T \quad \text{(at constant volume)}$$

The **equation of state of an ideal gas** is shown in two forms: expressed in terms of the number of moles (n) and the universal gas constant (R) as in Equation (10.8) or in terms of the number of molecules (N) and the Boltzman constant (k_B) as in Equation (10.11). *The temperature T must always be expressed in kelvins.*

$$PV = nRT \qquad (10.8)$$

n = number of moles

$R = 8.31 \text{ J/mol} \cdot \text{K} = 0.0821 \text{ L} \cdot \text{atm/mol} \cdot \text{K}$

$$PV = Nk_B T \qquad (10.11)$$

N = number of molecules

$$k_B = \frac{R}{N_A} = 1.38 \times 10^{-23} \text{ J/K} \qquad (10.12)$$

The **number of moles** in a sample of gas equals the number of molecules, N, divided by Avogadro's number, N_A.

$$n = \frac{N}{N_A} \qquad (10.10)$$

The **pressure of an ideal gas** is proportional to the number of molecules per unit volume and to the average kinetic energy of the molecules.

$$P = \tfrac{2}{3}\left(\frac{N}{V}\right)\left(\tfrac{1}{2} m \overline{v^2}\right) \qquad (10.13)$$

The **average translational kinetic energy** per molecule is directly proportional to the absolute temperature of the gas.

$$\tfrac{1}{2} m \overline{v^2} = \tfrac{3}{2} k_B T \qquad (10.15)$$

The **root-mean-square (rms) speed** expression shows that, at a given temperature, lighter molecules move faster on the average than heavier ones.

$$v_{\text{rms}} = \sqrt{\overline{v^2}} = \sqrt{\frac{3k_B T}{m}} = \sqrt{\frac{3RT}{M}} \qquad (10.18)$$

m = molecular mass

M = molar mass in kg/mole

REVIEW CHECKLIST

• Describe the operation of the constant-volume gas thermometer and how it is used to determine the Kelvin temperature scale. Convert between the various temperature scales, especially the conversion from degrees Celsius into kelvins, degrees Fahrenheit into kelvins, and degrees Celsius into degrees Fahrenheit.

• Define the linear expansion coefficient and volume expansion coefficient for an isotropic solid, and understand how to use these coefficients in practical situations involving expansion or contraction.

- Understand the assumptions made in developing the molecular model of an ideal gas; and apply the equation of state for an ideal gas to calculate pressure, volume, temperature, or number of moles.

- Define each of the following terms: molecular weight, mole, Avogadro's number, universal gas constant, and Boltzman's constant.

SOLUTIONS TO SELECTED END-OF-CHAPTER PROBLEMS

1. For each of the following temperatures, find the equivalent temperature on the indicated scale: (a) $-273.15°C$ on the Fahrenheit scale, (b) $98.6°F$ on the Celsius scale, and (c) 100 K on the Fahrenheit scale.

Solution

(a) Note that the given temperature, $T = -273.15°C = 0$ K, corresponds to absolute zero, the temperature at which an ideal gas will exert zero pressure. To convert this to the Fahrenheit scale, we use $T_F = \frac{9}{5}T_C + 32$ and find that absolute zero on the Fahrenheit scale occurs at

$$T_F = \frac{9}{5}(-273.15) + 32.00 = -459.67°F \approx -460°F \qquad \Diamond$$

(b) In this case, the given temperature, $T_F = 98.6°F$, is normal body temperature. On the Celsius scale, this temperature is

$$T_C = \frac{5}{9}(T_F - 32) = \frac{5}{9}(98.6 - 32.0) = 37.0°C \qquad \Diamond$$

(c) To convert a temperature on the Kelvin scale to the Fahrenheit scale, we first convert to the Celsius scale. Then, we convert the temperature from the Celsius to the Fahrenheit scale.

If $T = 100$ K, the corresponding temperature on the Celsius scale is

$$T_C = T - 273.15 = 100 - 273.15 = -173.15°C$$

Then, the corresponding temperature on the Fahrenheit scale is found to be

$$T_F = \frac{9}{5}(-173.15) + 32.00 = -279.67°F \approx -280°F \qquad \Diamond$$

5. Show that the temperature $-40°$ is unique in that it has the same numerical value on the Celsius and Fahrenheit scales.

Solution

The conversion from Celsius temperatures to the corresponding Fahrenheit temperature is $T_F = \frac{9}{5}T_C + 32$.

Observe that for a fairly high Celsius temperature, such as $T_C = 100°C$, the corresponding Fahrenheit temperature is $T_F = \frac{9}{5}(100) + 32 = 212°F$ or the numeric value of the Fahrenheit scale reading is greater than the corresponding reading on the Celsius scale. On the other hand, if the Celsius temperature is quite low, such as $T_C = -100°C$, the corresponding Fahrenheit temperature is found to be $T_F = \frac{9}{5}(-100) + 32 = -148°F$ or the Fahrenheit scale reading is lower than the corresponding reading on the Celsius scale. Thus, we see that the graphs of the readings on these two temperature scales vs. the Kelvin scale must cross one another, and there must be some temperature at which the two temperature scales will read the same.

To find the temperature which has the same numeric value on both the Celsius and the Fahrenheit scale, we substitute $T_C = T_F$ into the conversion equation. This gives

$$T_F = \frac{9}{5}T_F + 32 \qquad \text{or} \qquad -\frac{4}{5}T_F = 32$$

and yields

$$T_F = -\frac{5}{4}(32) = -40° \qquad\qquad\qquad \Diamond$$

as the single temperature at which the readings on the two scales have the same numeric value.

13. A pair of eyeglass frames are made of epoxy plastic (coefficient of linear expansion = 1.30×10^{-4} °C^{-1}). At room temperature (20.0°C), the frames have circular lens holes 2.20 cm in radius. To what temperature must the frames be heated if lenses 2.21 cm in radius are to be inserted into them?

Solution

There are two approaches one could take in the solution of this problem. One is to realize that, at 20.0°C, the epoxy material forming the circumference of one of the holes in the frames is shorter than the circumference of the lens. Then, the problem is to find the temperature to which the epoxy frames should be heated so the original circumference of the hole $C_0 = 2\pi r_0 = 2\pi(2.20 \text{ cm})$ will have expanded to match the circumference $C = 2\pi r = 2\pi(2.21 \text{ cm})$ of the lens. Taking this approach and using the linear expansion equation, we have $C = C_0[1 + \alpha(\Delta T)]$, or

$$2\pi(2.21 \text{ cm}) = 2\pi(2.20 \text{ cm})\left[1 + \left(1.30 \times 10^{-4} \text{ °C}^{-1}\right)(T - 20.0°C)\right] \qquad (1)$$

giving

$$T - 20.0°C = \frac{1}{1.30 \times 10^{-4} \text{ °C}^{-1}} \cdot \left[\frac{2\pi(2.21 \text{ cm})}{2\pi(2.20 \text{ cm})} - 1\right] = 35.0°C$$

and the desired temperature is

$$T = 35.0°C + 20.0°C = 55.0°C \qquad \qquad \Diamond$$

The other way one could think of this problem is to consider the expansion of the radius of the hole, realizing that linear dimensions of a hole or void in a surrounding material expand exactly the same as they would if filled with the surrounding material. That is: When the epoxy material is heated, the circular hole in that material expands exactly the same as a circular disk of epoxy material having the same radius as the hole (perhaps the disk that was cut away to create the hole) would expand. If we take this approach, we are computing the temperature to which the epoxy material should be heated so the original radius of the hole will match the radius of the lens. Then, we would write the linear expansion equation (using the expansion coefficient of the epoxy) as $r = r_0[1 + \alpha(\Delta T)]$, or

$$(2.21 \text{ cm}) = (2.20 \text{ cm})\left[1 + \left(1.30 \times 10^{-4} \text{ °C}^{-1}\right)(T - 20.0°C)\right]$$

Observe that this is the same as equation (1) above with the factors of 2π canceled. Thus, the two approaches are fully equivalent.

17. Lead has a density of 11.3×10^3 kg/m^3 at 0°C. (a) What is the density of lead at 90°C? (b) Based on your answer to part (a), now consider a situation in which you plan to invest in a gold bar. Would you be better off buying it on a warm day? Explain.

Solution

(a) As the temperature of a material rises, the mass of the material present does not change, however its volume increases. Therefore, the density (the mass per unit volume) will decrease with increasing temperature. The original density will be $\rho_0 = m/V_0$, where m is the mass of material present and V_0 is the volume occupied at the original temperature.

At the new temperature, the density is $\rho = m/V$, where the new volume is $V = V_0 + \Delta V$. The change in volume ΔV that occurs with change in temperature $\Delta T = 90.0°C - 0°C = 90.0°C$ is given by the volume expansion equation as $\Delta V = \beta_{lead} V_0 (\Delta T)$ with, $\beta_{lead} = 3\alpha_{lead} = 3(29 \times 10^{-6} \ °C^{-1}) = 87 \times 10^{-6} \ °C^{-1}$, so

$$V = V_0 + \beta_{lead} V_0 (\Delta T) = V_0 \left[1 + \beta_{lead} (\Delta T)\right]$$

and

$$\rho = \frac{m}{V_0 \left[1 + \beta_{lead} (\Delta T)\right]} = \frac{\rho_0}{1 + \beta_{lead} (\Delta T)}$$

At $\Delta T = 90.0°C - 0°C = 90.0°C$,

$$\rho_{lead} = \frac{11.3 \times 10^3 \ \text{kg/m}^3}{1 + (87 \times 10^{-6} \ °C^{-1})(90.0°C)} = 11.2 \times 10^3 \ \text{kg/m}^3 \qquad \Diamond$$

(b) You would not be better off buying the gold bar on a warm day. While the density of gold would be less on a warm day, the mass of the bar would be the same—and that is what you are paying for. As the temperature rises, the volume increases, causing the density to decrease, but the mass remains constant.

---

25. The average coefficient of volume expansion for carbon tetrachloride is $5.81 \times 10^{-4} \ (°C)^{-1}$. If a 50.0-gal steel container is filled completely with carbon tetrachloride when the temperature is 10.0°C, how much will spill over when the temperature rises to 30.0°C?

Solution

As the temperature rises, both the carbon tetrachloride and the steel container will expand. However, the volume expansion coefficient for carbon tetrachloride, $\beta_{C(Cl)_4} = 5.81 \times 10^{-4} \ (°C)^{-1}$, is greater than that of steel, $\beta_{steel} = 3\alpha_{steel} = 33.0 \times 10^{-6} \ (°C)^{-1}$. Thus, the carbon tetrachloride expands faster than the steel container, and since the container was completely filled at the original temperature, spillage will occur as the temperature rises.

The amount of spillage that occurs is the amount by which the expansion of the carbon tetrachloride volume exceeds the expansion of the steel container's volume. That is,

$$V_{spillage} = \Delta V_{C(Cl_4)} - \Delta V_{steel \atop container} = \beta_{C(Cl_4)} V_0 (\Delta T) - \beta_{steel} V_0 (\Delta T) = V_0 \left(\beta_{C(Cl_4)} - \beta_{steel} \right)(\Delta T)$$

Note that the original volume, $V_0 = 50.0$ gal, and the change in temperature, $\Delta T = +20.0°C$, are the same for both the carbon tetrachloride and the steel container. Also, note that we are treating the steel container as though it were solid steel, rather than a steel surface around a cavity. Please see the solution of Problem 13 earlier in this chapter for a discussion of this concept as it relates to linear expansion.

We then have

$$V_{spillage} = (50.0 \text{ gal}) \left[5.81 \times 10^{-4} \ (°C)^{-1} - 33.0 \times 10^{-6} \ (°C)^{-1} \right] (20.0° \text{ C})$$

or

$$V_{spillage} = (50.0 \text{ gal}) \left[5.81 \times 10^{-4} \ (°C)^{-1} - 33.0 \times 10^{-6} \ (°C)^{-1} \right] (20.0° \text{ C}) = 0.548 \text{ gal} \qquad \Diamond$$

---

29. One mole of oxygen gas is at a pressure of 6.00 atm and a temperature of 27.0°C. (a) If the gas is heated at constant volume until the pressure triples, what is the final temperature? (b) If the gas is heated so that both the pressure and volume are doubled, what is the final temperature?

Solution

(a) We shall treat the oxygen as an ideal gas and apply the ideal gas law, $PV = nRT$. When n and V are both held constant, we see that

$$\frac{P_f V_f}{P_i V_i} = \frac{n_f R T_f}{n_i R T_i}$$

or

$$T_f = T_i \left(\frac{P_f}{P_i} \right)$$

Remembering to always use absolute temperatures with the ideal gas law

$$(T_i = 27.0 + 273.15 = 300 \text{ K})$$

this yields

$$T_f = (300 \text{ K}) \left(\frac{3P_i}{P_i} \right) = 900 \text{ K}$$

or

$$(T_C)_f = 900 - 273.15 = 627°C \qquad \Diamond$$

(b) Again using the ideal gas law, with only n and R constant,

$$\frac{P_f V_f}{P_i V_i} = \frac{n_f R T_f}{n_i R T_i}$$

or

$$\frac{T_f}{T_i} = \frac{P_f V_f}{P_i V_i} = \frac{(2P_i)(2V_i)}{P_i V_i} = 4$$

Thus,

$$T_f = 4T_i = 4(300 \text{ K}) = 1\,200 \text{ K}$$

or

$$(T_C)_f = 1\,200 - 273.15 = 927°C \qquad \Diamond$$

---

35. A weather balloon is designed to expand to a maximum radius of 20 m at its working altitude, where the air pressure is 0.030 atm and the temperature is 200 K. If the balloon is filled at atmospheric pressure and 300 K, what is its radius at liftoff?

Solution

We neglect any effects due to elastic forces in the material making up the skin of the balloon, and assume that the gas pressure inside the balloon always equals the atmospheric pressure outside the balloon. Then, the thermodynamic properties of the gas at liftoff are

$$P_i = 1.0 \text{ atm}, \ T_i = 300 \text{ K}, \text{ and } V_i = 4\pi r_i^3/3$$

The properties of the gas when the balloon is at the working altitude are

$$P_f = 0.030 \text{ atm}, \ T_f = 200 \text{ K}, \text{ and } V_f = 4\pi r_f^3/3 = 4\pi(20 \text{ m})^3/3$$

Using the ideal gas law and recognizing that $n_f = n_i$ (assuming no leakage as the balloon ascends), we obtain

$$\frac{P_i V_i}{P_f V_f} = \frac{n_i R T_i}{n_f R T_f}$$

or

$$V_i = V_f\left(\frac{P_f}{P_i}\right)\left(\frac{T_i}{T_f}\right) \quad \text{and} \quad \left(\frac{4\pi}{3}\right)r_i^3 = \left(\frac{4\pi}{3}\right)(20 \text{ m})^3\left(\frac{P_f}{P_i}\right)\left(\frac{T_i}{T_f}\right)$$

Thus,

$$r_i = (20 \text{ m})\left[\left(\frac{0.030 \text{ atm}}{1.0 \text{ atm}}\right)\left(\frac{300 \text{ K}}{200 \text{ K}}\right)\right]^{\frac{1}{3}} = 7.1 \text{ m} \qquad \Diamond$$

---

41. Use Avogadro's number to find the mass of a helium atom.

Solution

One mole of any substance contains Avogadro's number N_A of molecules and has a mass equal to the molar mass or molecular weight M of that substance. Thus, if $m_{molecule}$ is the mass of a single molecule of the substance, we have that

$$N_A m_{molecule} = M$$

or

$$m_{molecule} = \frac{M}{N_A}$$

Helium, an inert gas, is a monatomic substance. This means that each molecule of helium contains a single helium atom, so the mass of a helium molecule and that of a helium atom are identical. Since the molecular weight of helium is $M_{He} = 4.00$ g/mol, and Avogadro's number is $N_A = 6.02 \times 10^{23}$ molecules/mol, we have for helium

$$m_{molecule} = \frac{M_{He}}{N_A} = \frac{4.00 \text{ g/mol}}{6.02 \times 10^{23} \text{ molecules/mol}} = 6.64 \times 10^{-24} \text{ g/molecule}$$

and

$$m_{atom} = m_{molecule} = 6.64 \times 10^{-24} \text{ g/atom} \qquad \Diamond$$

44. A 7.00-L vessel contains 3.50 moles of ideal gas at a pressure of 1.60×10^6 Pa. Find (a) the temperature of the gas and (b) the average kinetic energy of a gas molecule in the vessel. (c) What additional information would you need if you were asked to find the average speed of a gas molecule?

Solution

(a) The volume the gas occupies is

$$V = 7.00 \text{ L} = \left(7.00 \text{ L}\right)\left(\frac{10^3 \text{ cm}^3}{1 \text{ L}}\right)\left(\frac{1 \text{ m}^3}{10^6 \text{ cm}^3}\right) = 7.00 \times 10^{-3} \text{ m}^3$$

The ideal gas law, $PV = nRT$, then gives the absolute temperature as

$$T = \frac{PV}{nR} = \frac{\left(1.60 \times 10^6 \text{ Pa}\right)\left(7.00 \times 10^{-3} \text{ m}^3\right)}{(3.50 \text{ mol})(8.31 \text{ J/mol} \cdot \text{K})} = 385 \text{ K} \qquad \Diamond$$

(b) From kinetic theory, the average kinetic energy of a molecule in an ideal gas (or any gas that approximates an ideal gas) is

$$\overline{KE}_{molecule} = \frac{1}{2}m\overline{v^2} = \frac{3}{2}k_BT$$

where k_B is Boltzmann's constant and T is the absolute temperature of the gas. Thus, for the given gas, we have

$$\overline{KE}_{molecule} = \frac{3}{2}\left(1.38\times10^{-23}\ \text{J/K}\right)(385\ \text{K}) = 7.97\times10^{-21}\ \text{J} \qquad \Diamond$$

(c) The average molecular speed (actually the rms, or root-mean-square speed) can be found from

$$\overline{KE}_{molecule} = m\overline{v^2}/2 \qquad \text{as} \qquad v_{rms} = \sqrt{\overline{v^2}} = \sqrt{2\overline{KE}_{molecule}/m}$$

To carry out this computation, one needs the mass m of a single gas molecule, which in turn requires a knowledge of the molecular weight of the gas (see the solution of Problem 41 earlier in this chapter.) $\qquad \Diamond$

---

49. A popular brand of cola contains 6.50 g of carbon dioxide dissolved in 1.00 L of soft drink. If the evaporating carbon dioxide is trapped in a cylinder at 1.00 atm and 20.0°C, what volume does the gas occupy?

Solution

The molecular weight of carbon dioxide (CO_2) is

$$M_{CO_2} = (\text{atomic mass of carbon}) + 2(\text{atomic mass of oxygen})$$
$$= [12.0 + 2(16.0)]\ \text{g/mol} = 44.0\ \text{g/mol}$$

The number of moles of carbon dioxide dissolved in the 1.00-L of soft drink is

$$n = \frac{m}{M_{CO_2}} = \frac{6.50\ \text{g}}{44.0\ \text{g/mol}} = 0.148\ \text{mol}$$

Under the specified conditions,

$$P = 1.00\ \text{atm} = 1.013\times10^5\ \text{Pa},\ T = 20°\text{C} = 293\ \text{K}$$

the ideal gas law gives the volume occupied by this quantity of gas as

$$V = \frac{nRT}{P} = \frac{(0.148\ \text{mol})(8.31\ \text{J/mol}\cdot\text{K})(293\ \text{K})}{1.013\times10^5\ \text{Pa}}$$

or

$$V = 3.55\times10^{-3}\ \text{m}^3 = \left(3.55\times10^{-3}\ \text{m}^3\right)\left(\frac{10^6\ \text{cm}^3}{1\ \text{m}^3}\right)\left(\frac{1\ \text{L}}{10^3\ \text{cm}^3}\right) = 3.55\ \text{L} \qquad \Diamond$$

---

58. Before beginning a long trip on a hot day, a driver inflates an automobile tire to a gauge pressure of 1.80 atm at 300 K. At the end of the trip the gauge pressure has increased to 2.20 atm. (a) Assuming that the volume has remained constant, what is the temperature of the air inside the tire? (b) What percentage of the original mass of air in the tire should be released so the pressure returns to its original value? Assume that the temperature remains at the value found in (a) and the volume of the tire remains constant as air is released.

Solution

(a) It is important to remember that one must always use the absolute pressure and the absolute temperature when applying the ideal gas law. The initial and final absolute pressures of the air in the tire are

$$P_i = P_{atm} + \left(P_{gauge}\right)_i = 1.00 \text{ atm} + 1.80 \text{ atm} = 2.80 \text{ atm}$$

and

$$P_f = P_{atm} + \left(P_{gauge}\right)_f = 1.00 \text{ atm} + 2.20 \text{ atm} = 3.20 \text{ atm}$$

Thus, with volume and quantity of gas both constant, the ideal gas law gives

$$\frac{P_f V_f}{P_i V_i} = \frac{n_f R T_f}{n_i R T_i}$$

or

$$T_f = T_i\left(\frac{P_f}{P_i}\right) = (300 \text{ K})\left(\frac{3.20 \text{ atm}}{2.80 \text{ atm}}\right) = 343 \text{ K} \qquad \Diamond$$

(b) From the ideal gas law, $PV = nRT$, the number of moles of gas present is $n = PV/RT$. Thus, if the volume enclosed by the tire and the temperature of the gas in the tire remain constant as gas is released, we have

$$\frac{n_f}{n_i} = \frac{P_f V / RT}{P_i V / RT} = \frac{P_f}{P_i} = \frac{2.80 \text{ atm}}{3.20 \text{ atm}} = 0.875$$

or 87.5% of the number of moles originally in the tire is still present after the pressure is adjusted. Thus, 12.5% of the gas originally in the tire should be released.

Since the mass of the gas is directly proportional to the number of moles present, 12.5% of the original mass of air in the tire should be released. $\qquad \Diamond$

64. Two small containers, each with a volume of 100 cm^3, contain helium gas at 0°C and 1.00 atm pressure. The two containers are joined by a small open tube of negligible volume, allowing gas to flow from one container to the other. What common pressure will exist in the two containers if the temperature of one container is raised to 100°C while the other container is kept at 0°C?

Solution

We shall call the container that is kept at 0°C container 1, and the one whose temperature is raised to 100°C will be container 2.

As the temperature of container 2 is raised, gas may flow from one container to the other, but no gas escapes. Thus, at the end, the total number of moles of gas in the joined containers is the same as at the beginning, or

$$n_{1f} + n_{2f} = n_{1i} + n_{2i}$$

From the ideal gas law, $PV = nRT$, we have $n = PV/RT$ and our previous result becomes

$$\frac{P_{1f}V_{1f}}{RT_{1f}} + \frac{P_{2f}V_{2f}}{RT_{2f}} = \frac{P_{1i}V_{1i}}{RT_{1i}} + \frac{P_{2i}V_{2i}}{RT_{2i}}$$

Since the two containers have equal initial volumes, and ignoring the very slight change in the volume of container 2 as its temperature is raised, we have $V_{1i} = V_{2i} = V_{1f} = V_{2f}$ and all volumes cancel in the above equation. Also, the two containers are joined, maintaining equal pressures in them at all times. Thus, $P_{1f} = P_{2f} = P_f$ and $P_{1i} = P_{2i} = 1.00$ atm. With these values, our previous result becomes

$$P_f \left[\frac{\cancel{V_{1f}}}{\cancel{R}T_{1f}} + \frac{\cancel{V_{2f}}}{\cancel{R}T_{2f}} \right] = (1.00 \text{ atm}) \left[\frac{\cancel{V_{1i}}}{\cancel{R}T_{1i}} + \frac{\cancel{V_{2i}}}{\cancel{R}T_{2i}} \right]$$

or

$$P_f = (1.00 \text{ atm}) \left(\frac{T_{2i} + T_{1i}}{T_{1i}T_{2i}} \right) \left(\frac{T_{1f}T_{2f}}{T_{2f} + T_{1f}} \right)$$

Finally, we recognize that the absolute temperatures are $T_{1i} = T_{2i} = T_{1f} = 0°C = 273$ K and $T_{2f} = 100°C = 373$ K, so we have

$$P_f = (1.00 \text{ atm}) \left[\frac{273 \text{ K} + 273 \text{ K}}{(273 \text{ K})\cancel{(273 \text{ K})}} \right] \left[\frac{\cancel{(273 \text{ K})}(373 \text{ K})}{373 \text{ K} + 273 \text{ K}} \right] = 1.15 \text{ atm} \qquad \lozenge$$

Observe that, in order to solve this problem, it was not necessary to know the actual volumes of the containers, only that it was very large in comparison to any increase in the volume of container 2 as it was heated. Also, it was not necessary to know the specific gas in the container, only that it behaved like an ideal gas.

11

Energy in Thermal Processes

NOTES FROM SELECTED CHAPTER SECTIONS

11.1 Heat and Internal Energy

When two systems at different temperatures are placed in thermal contact, energy is transferred by heat from the warmer to the cooler object until they reach a common temperature (i.e., when they are in thermal equilibrium with each other). Heat is a mechanism by which energy (measured in calories or joules) is transferred between a system and its environment due to a temperature difference.

The **mechanical equivalent of heat,** first measured by Joule, is given by $1 \text{ cal} = 4.186$ J. This is the definition of the calorie.

11.2 Specific Heat

The **specific heat**, c, of a substance is the energy per unit mass per degree Celsius temperature increase for that substance.

11.3 Calorimetry

Calorimetry is a procedure carried out in an isolated system (usually a substance of unknown specific heat and water) in which the transfer of thermal energy occurs. A negligible amount of mechanical work is done in the process. The law of conservation of energy in a calorimeter requires that the energy that leaves the warmer substance (of unknown specific heat) equals the energy that enters the water.

11.4 Latent Heat and Phase Change

A substance usually undergoes a change in temperature when energy is transferred by heat between it and its surroundings. There are situations, however, in which the transfer of energy does not result in a change in temperature. This is the case whenever the substance undergoes a **phase change.** Some common phase changes are solid to liquid, liquid to gas, and a change in crystalline structure of a solid. *Every phase change involves a change in internal energy without an accompanying change in temperature.*

The **latent heat of fusion** is a parameter used to characterize a solid-to-liquid phase change; the **latent heat of vaporization** characterizes the liquid-to-gas phase change.

11.5 Energy Transfer
There are three basic processes of thermal energy transfer: conduction, convection, and radiation.

Conduction is an energy transfer process that occurs in a substance when there is a temperature gradient across the substance. That is, conduction of energy occurs only when the temperature of the substance is **not** uniform. For example, if there is a temperature difference between the two ends of a metal rod, energy will flow from the hot end to the colder end. The rate of flow of heat along the rod is proportional to the cross-sectional area of the rod, the temperature gradient, and k, the thermal conductivity of the material of which the rod is made.

Convection is a process of energy transfer by the motion of material, such as the mixing of hot and cold fluids. Convection heating is used in conventional hot-air and hot-water heating systems. Convection currents produce changes in weather conditions when warm and cold air masses mix in the atmosphere.

Radiation is energy transfer by the emission or absorption of electromagnetic waves.

EQUATIONS AND CONCEPTS

The **mechanical equivalent of heat** was first measured by Joule. Equation (11.1) is the definition of the calorie as an energy unit.

$$1 \text{ cal} = 4.186 \text{ J} \tag{11.1}$$

The **specific heat,** c, of a substance is the energy per unit mass per degree Celsius temperature increase for that substance. *In this and subsequent equtions, Q represents the quantity of energy transferred by heat between a system and its environment. Specific heat is characteristic of a **specific type** of material.*

$$c = \frac{Q}{m\Delta T} \tag{11.2}$$

SI units of $c = \text{J/kg} \cdot {}^\circ\text{C}$

Calorimetry, a technique to measure specific heat, is based on conservation of energy in an isolated system. *The negative sign is required so that both sides of Equation (11.4) will be positive.*

$$Q_{cold} = -Q_{hot} \tag{11.4}$$

Q_{cold} is a positive quantity.

Q_{hot} is a negative quantity.

Latent heat (L) Is a thermal property of a substance that determines the quantity of energy required to cause a mass m of that substance to undergo a phase change. The latent heat of fusion, L_f is used when the phase change is from solid to liquid (or liquid to solid). The latent heat of vaporization, L_v is used when the phase change is from liquid to gas (or gas to liquid). *A phase change occurs at constant temperature and is accompanied by a change in internal energy.*

$$Q = \pm mL \tag{11.6}$$

Use + when energy enters a system.

Use − when energy leaves a system.

L is expressed in units of J/kg.

Energy transfer by conduction through a slab or rod of material is directly proportional to the temperature difference between the hot and cold faces and to the cross-sectional area of the slab and indirectly proportional to the thickness of the slab (or the length of the rod). *The thermal conductivity parameter is characteristic of a particular material.*

$$\mathcal{P} = kA\left(\frac{T_h - T_c}{L}\right) \tag{11.7}$$

k = thermal conductivity

Energy transfer through a compound slab is accomplished by a summation over all layers of the slab. For this calculation, T_h and T_c are the temperatures of the outer extremities of the slab. The thicknesses of the layers of the slab are $L_1, L_2, L_3, \ldots$, and $k_1, k_2, k_3, \ldots$ are respective thermal conductivities. The rate of energy transfer, can also be expressed in cal/s and Btu/h.

$$\frac{Q}{\Delta t} = \frac{A(T_h - T_c)}{\sum_i L_i/k_i} \tag{11.8}$$

$$\text{or } \frac{Q}{\Delta t} = \frac{A(T_h - T_c)}{\sum_i R_i} \tag{11.9}$$

where $R_i = \dfrac{L_i}{k_i}$

Stefan's law expresses the **rate of energy transfer by radiation** (radiated power). *The radiated power is proportional to the fourth power of the absolute temperature.*

$$\mathcal{P} = \sigma A e T^4 \tag{11.10}$$

$\sigma = 5.670 \times 10^{-8} \ \text{W/m}^2 \cdot \text{K}^4$

T = surface temperature in kelvins

e = emissivity (values from 0 to 1)

Net radiated power by an object at temperature T in an environment at temperature T_0 is the difference between radiated and absorbed power. *At thermal equilibrium ($T = T_0$), an object radiates and absorbs energy at the same rate and the temperature of the object remains constant.*

$$\mathcal{P}_{\text{net}} = \sigma A e\left(T^4 - T_0^{\ 4}\right) \tag{11.11}$$

SUGGESTIONS, SKILLS, AND STRATEGIES

If you are having difficulty with calorimetry problems, one or more of the following factors should be considered:

1. Be sure your units are consistent throughout. That is, if you are using specific heats measured in J/kg·°C, be sure that masses are in kilograms and temperatures are in Celsius units throughout.

2. Losses and gains in energy are found by using $Q = mc\Delta T$ only for those intervals in which no phase changes occur. Likewise, the equations $Q = \pm m L_f$ and $Q = \pm m L_v$ are to be used only when phase changes are taking place.

3. Often sign errors occur applying the basic calorimetry equation ($Q_{cold} = -Q_{hot}$). Remember to include the negative sign in the equation; and remember that ΔT **is always the final temperature minus the initial temperature.**

REVIEW CHECKLIST

- Define and discuss the calorie, Btu, specific heat, and latent heat. Convert among calories, Btu's, and joules.

- Use equations for specific heat, latent heat, temperature change, and energy gain (loss) to solve calorimetry problems.

- Discuss the possible mechanisms which can give rise to energy transfer between a system and its surroundings (i.e., conduction, convection, and radiation), and give a realistic example of each transfer mechanism.

- Apply the basic law of thermal conduction, and Stefan's law for energy transfer by radiation.

SOLUTIONS TO SELECTED END-OF-CHAPTER PROBLEMS

3. Lake Erie contains roughly $4.00 \times 10^{11} \text{ m}^3$ of water. (a) How much energy is required to raise the temperature of that volume of water from 11.0°C to 12.0°C? (b) How many years would it take to supply this amount of energy by using the 1 000-MW exhaust energy of an electric power plant?

Solution

(a) The mass of the water in Lake Erie is

$$m = \rho V = \left(1.00 \times 10^3 \text{ kg/m}^3\right)\left(4.00 \times 10^{11} \text{ m}^3\right) = 4.00 \times 10^{14} \text{ kg}$$

and the energy needed to raise the temperature of this water from 11.0°C to 12.0°C is given by $Q = mc(\Delta T)$ as

$$Q = \left(4.00 \times 10^{14} \text{ kg}\right)\left(4\,186 \text{ J/kg}\cdot°\text{C}\right)\left(12.0°\text{C} - 11.0°\text{C}\right) = 1.67 \times 10^{18} \text{ J} \qquad \lozenge$$

(b) The exhaust from the power plant would add energy to the water of Lake Erie at a rate of

$$\mathcal{P} = 1\,000 \text{ MW} = 1\,000 \times 10^6 \text{ W} = 1.00 \times 10^9 \text{ J/s}$$

and the time needed to raise the temperature of the lake from 11.0°C to 12.0°C is

$$t = \frac{Q}{\mathcal{P}} = \frac{1.67 \times 10^{18} \text{ J}}{1.00 \times 10^9 \text{ J/s}} = 1.67 \times 10^9 \text{ s} \left(\frac{1 \text{ yr}}{3.156 \times 10^7 \text{ s}} \right) = 52.9 \text{ yr} \qquad \lozenge$$

---

11. A 200-g aluminum cup contains 800 g of water in thermal equilibrium with the cup at 80°C. The combination of cup and water is cooled uniformly so that the temperature decreases by 1.5°C per minute. At what rate is energy being removed? Express your answer in watts.

Solution

Note that in the temperature range involved here, the water in the cup will be in the liquid state, while the aluminum of the cup will be in the solid state. Consider a 1 minute time interval during which the temperature of the system consisting of the 200-g aluminum cup and 800 g of water undergoes a change in temperature of $\Delta T = -1.5°C$. The thermal energy that must be removed from the system to accomplish this change in temperature is

$$|Q| = m_{cup} c_{Al} |\Delta T| + m_{water} c_{\substack{liquid \\ water}} |\Delta T|$$

and using the values of specific heat from Table 11.1 of the textbook,

$$|Q| = (0.200 \text{ kg})(900 \text{ J/kg} \cdot °C)(1.5°C) + (0.800 \text{ kg})(4\,186 \text{ J/kg} \cdot °C)(1.5°C)$$

yielding $|Q| = 5.3 \times 10^3$ J. The rate at which energy is being removed is then

$$\mathcal{P} = \frac{|Q|}{\Delta t} = \frac{5.3 \times 10^3 \text{ J}}{1.0 \text{ min}} \left(\frac{1 \text{ min}}{60 \text{ s}} \right) = 88 \text{ J/s} = \boxed{88 \text{ W}} \qquad \lozenge$$

17. An aluminum cup contains 225 g of water and a 40-g copper stirrer all at 27°C. A 400-g sample of silver at an initial temperature of 87°C is placed in the water. The stirrer is used to stir the mixture until it reaches its final equilibrium temperature of 32°C. Calculate the mass of the aluminum cup.

Solution

Assuming that the system consisting of the cup, stirrer, water, and silver sample is isolated from the environment, the total energy content of the system is constant. Therefore, $Q_{cold} = -Q_{hot}$, or the energy gained by the initially cooler parts (cup, stirrer, and water) equals the energy lost by the initially warmer parts (the silver sample):

$$Q_{cup} + Q_{stirrer} + Q_{water} = -Q_{Ag}$$

Since the cup, stirrer, and water all undergo the same change in temperature, this becomes

$$\left[m_{cup} c_{Al} + m_{stirrer} c_{Cu} + m_{water} c_{water} \right] (\Delta T_{water}) = -m_{Ag} c_{Ag} (\Delta T_{Ag})$$

The mass of the cup is then

$$m_{cup} = m_{Ag} \left(\frac{c_{Ag}}{c_{Al}} \right) \left[\frac{-\Delta T_{Ag}}{\Delta T_{water}} \right] - m_{stirrer} \left(\frac{c_{Cu}}{c_{Al}} \right) - m_{water} \left(\frac{c_{water}}{c_{Al}} \right)$$

or

$$m_{cup} = (400 \text{ g}) \left(\frac{234 \text{ J/kg} \cdot {}^{\circ}\text{C}}{900 \text{ J/kg} \cdot {}^{\circ}\text{C}} \right) \left[\frac{-(32^{\circ}\text{C} - 87^{\circ}\text{C})}{32^{\circ}\text{C} - 27^{\circ}\text{C}} \right]$$

$$- (40 \text{ g}) \left(\frac{387 \text{ J/kg} \cdot {}^{\circ}\text{C}}{900 \text{ J/kg} \cdot {}^{\circ}\text{C}} \right) - (225 \text{ g}) \left(\frac{4\,186 \text{ J/kg} \cdot {}^{\circ}\text{C}}{900 \text{ J/kg} \cdot {}^{\circ}\text{C}} \right)$$

giving

$$m_{cup} = 80 \text{ g}$$ ◊

24. An unknown substance has a mass of 0.125 kg and an initial temperature of 95.0°C. The substance is then dropped into a calorimeter made of aluminum containing 0.285 kg of water initially at 25.0°C. The mass of the aluminum container is 0.150 kg, and the temperature of the calorimeter increases to a final equilibrium temperature of 32.0°C. Assuming no thermal energy is transferred to the environment, calculate the specific heat of the unknown substance.

Solution

If no thermal energy is transferred to or from the environment, any energy gained by the initially cooler materials must equal the energy given up by the initially warmer materials as the system approaches the equilibrium temperature. That is, $Q_{cold} = -Q_{hot}$, or

$$m_{Al}c_{Al}\left(T_f - T_{i,Al}\right) + m_{water}c_{water}\left(T_f - T_{i,water}\right) = -m_x c_x\left(T_f - T_{i,x}\right)$$

where the subscript Al refers to the aluminum calorimeter cup and the subscript x refers to the unknown substance. Observe that $T_{i,Al} = T_{i,water} = T_{cold} = 25.0°C$. The specific heat of the unknown material is then

$$c_x = \frac{\left[m_{Al}c_{Al} + m_{water}c_{water}\right]\left(T_f - T_{cold}\right)}{-m_x\left(T_f - T_{i,x}\right)}$$

or

$$c_x = \frac{\left[(0.150\ \text{kg})(900\ \text{J/kg}\cdot°\text{C}) + (0.285\ \text{kg})(4\,186\ \text{J/kg}\cdot°\text{C})\right](32.0°\text{C} - 25.0°\text{C})}{-(0.125\ \text{kg})(32.0°\text{C} - 95.0°\text{C})}$$

which yields

$$c_x = 1.18 \times 10^3\ \text{J/kg}\cdot°\text{C}$$ ◊

31. A 40-g block of ice is cooled to −78°C and is then added to 560 g of water in an 80-g copper calorimeter at a temperature of 25°C. Determine the final temperature of the system consisting of the ice, water, and calorimeter. (If not all the ice melts, determine how much ice is left.) Remember that the ice must first warm to 0°C, melt, and then continue warming as water. (The specific heat of ice is 0.500 cal/g · °C $= 2\,090$ J/kg · °C.)

Solution

The recommended first step in problems similar to this is to determine whether all of the ice will melt. To do this, we compute the minimum energy that must be added to the ice to melt all of it. This is $Q_{melt} = Q_{\substack{warm \\ ice\ to\ 0°C}} + Q_{\substack{ice\ to \\ liquid\ at\ 0°C}}$, or

$$Q_{melt} = m_{ice}c_{ice}\left(0°C - T_{i,ice}\right) + m_{ice}L_{f,water}$$

$$= (0.040\ \text{kg})\left[(2\,090\ \text{J/kg·°C})(78°C) + 3.33 \times 10^5\ \text{J/kg}\right.$$

giving $Q_{melt} = 2.0 \times 10^4$ J.

The energy the warmer materials can give up before their temperature reaches 0°C is

$$Q_{available} = -\left[m_{cal}c_{Cu} + m_{water}c_{water}\right]\left(0°C - T_{i,hot}\right)$$

or

$$Q_{available} = \left[(0.080\ \text{kg})(387\ \text{J/kg·°C}) + (0.560\ \text{kg})(4\,186\ \text{J/kg·°C})\right](25°C) = 5.9 \times 10^4\ \text{J}$$

and we see that the warmer materials can provide more than the minimum energy needed to melt the ice. The final temperature will be between 0°C and 25°C, so we do a standard calorimetry calculation, recognizing that the specific heat of the ice will be the same as that of liquid water after the ice has melted, to determine T_f. With the system thermally isolated from the environment, $Q_{cold} = -Q_{hot}$, or

$$m_{ice}\left\{c_{ice}\left[0°C - (-78°C)\right] + L_f + c_{water}\left(T_f - 0°C\right)\right\} = -\left(m_{water}c_{water} + m_{cal}c_{Cu}\right)\left(T_f - 25°C\right)$$

giving

$$T_f = \frac{\left(m_{water}c_{water} + m_{cal}c_{Cu}\right)(25°C) - m_{ice}\left[c_{ice}(78°C) + L_f\right]}{\left(m_{water} + m_{ice}\right)c_{water} + m_{cal}c_{Cu}}$$

With $m_{ice} = 0.040$ kg, $m_{cal} = 0.080$ kg, $m_{water} = 0.560$ kg, $L_f = 3.33 \times 10^5$ J/kg

$c_{ice} = 2\,090$ J/kg°C, $c_{Cu} = 387$ J/kg°C, and $c_{water} = 4\,186$ J/kg°C, we have

$$T_f = \frac{\left[(0.560)(4\,186) + (0.080)(387)\right](25°C) - (0.040)\left[(2\,090)(78°C) + 3.33 \times 10^5\right]}{(0.560 + 0.040)(4\,186) + 0.080(387)}$$

which yields

$$T_f = 16°C$$

◊

35. Steam at 100°C is added to ice at 0°C. (a) Find the amount of ice melted and the final temperature when the mass of steam is 10 g and the mass of ice is 50 g. (b) Repeat with steam of mass 1.0 g and ice of mass 50 g.

Solution

(a) With a fairly high ratio of steam to ice, we shall assume that all the ice will melt and the equilibrium temperature is somewhere above 0°C.

Equating the energy gained by the ice to the energy given up by the steam gives

$$m_{ice}L_f + m_{ice}c_{water}(\Delta T)_{\substack{melted \\ ice}} = m_{steam}L_v + m_{steam}c_{water}\left[-(\Delta T)_{\substack{condensed \\ steam}}\right]$$

or

$$(50 \text{ g})(79.7 \text{ cal/g}) + (50 \text{ g})(1.0 \text{ cal/g} \cdot °C)\left(T_f - 0°C\right) = (10 \text{ g})(540 \text{ cal/g})$$
$$+ (10 \text{ g})(1.0 \text{ cal/g} \cdot °C)\left[-\left(T_f - 100°C\right)\right]$$

Solving for the final temperature gives

$$T_f = \frac{5\,400 \text{ cal} + 1\,000 \text{ cal} - 4\,000 \text{ cal}}{50 \text{ cal/°C} + 10 \text{ cal/°C}} = 40°C \qquad \lozenge$$

(b) If we attempt to duplicate the previous calculation when only 1.0 g of steam is used, we obtain

$$(50 \text{ g})(79.7 \text{ cal/g}) + (50 \text{ g})(1.0 \text{ cal/g} \cdot °C)\left(T_f - 0°C\right) = (1.0 \text{ g})(540 \text{ cal/g})$$
$$+ (1.0 \text{ g})(1.0 \text{ cal/g} \cdot °C)\left[-\left(T_f - 100°C\right)\right]$$

and

$$T_f = \frac{540 \text{ cal} + 100 \text{ cal} - 4\,000 \text{ cal}}{50 \text{ cal/°C} + 1.0 \text{ cal/°C}} = -66°C$$

which is clearly incorrect. Adding steam to ice will not lower the temperature of the ice!

Carefully analyzing the previous solution shows that we assumed all the ice will melt. However, if only 1.0 g of steam is used, the total energy the steam can give up as it condenses and cools all the way to 0°C (the temperature of the ice) is

$$Q' = m_{steam}L_v + m_{steam}c_{water}\left[-(\Delta T)_{\substack{condensed \\ steam}}\right]$$
$$= (1.0 \text{ g})(540 \text{ cal/g}) + (1.0 \text{ g})(1.0 \text{ cal/g} \cdot °C)(100° C) = 640 \text{ cal}$$

The amount of ice this much energy can melt is only

$$m = \frac{Q'}{L_f} = \frac{640 \text{ cal}}{79.7 \text{ cal/g}} = 8.0 \text{ g}$$

Thus, we see that 42 grams of ice will remain and that all parts of the system (the remaining ice, the melted ice, and the condensed steam) come to an equilibrium temperature of 0°C in this case. $\qquad \lozenge$

41. A steam pipe is covered with 1.50-cm-thick insulating material of thermal conductivity 0.200 cal/cm·°C·s. How much energy is lost every second when the steam is at 200°C and the surrounding air is at 20.0°C? The pipe has a circumference of 800 cm and a length of 50.0 m. Neglect losses through the ends of the pipe.

Solution

The area of the inner surface of the insulation layer covering the cylindrical pipe is

$$A = (circumference) \times (length) = (800 \text{ cm}) \times (50.0 \times 10^2 \text{ cm}) = 4.00 \times 10^6 \text{ cm}^2$$

The rate of energy transfer by conduction through the insulation layer is given by the relation

$$\mathcal{P} = kA\left(\frac{T_h - T_c}{L}\right)$$

where k is the thermal conductivity of the material, A is the surface area of the conducting material, L is the thickness of the material, and $\Delta T = T_h - T_c$ is the difference in temperature on the two sides of the conducting material.

Thus, the rate of energy loss through the side of the steam pipe is

$$\mathcal{P} = \left(0.200 \ \frac{\text{cal}}{\text{cm} \cdot {}^\circ\text{C} \cdot \text{s}}\right)\left(4.00 \times 10^6 \text{ cm}^2\right)\left(\frac{200{}^\circ\text{C} - 20.0{}^\circ\text{C}}{1.50 \text{ cm}}\right) = 9.60 \times 10^7 \ \frac{\text{cal}}{\text{s}}$$

or

$$\mathcal{P} = 9.60 \times 10^7 \ \frac{\text{cal}}{\text{s}}\left(\frac{4.186 \text{ J}}{1 \text{ cal}}\right) = 4.02 \times 10^8 \ \text{J/s} = 4.02 \times 10^8 \ \text{W} = 402 \text{ MW} \qquad \lozenge$$

Clearly, this pipe needs to be better insulated.

48. A solar sail is made of aluminized Mylar having an emissivity of 0.03 and reflecting 97.0% of the light that falls on it. Suppose a sail with area 1.00 km^2 is oriented so that sunlight falls perpendicular to its surface with an intensity of $1.40 \times 10^3 \text{ W/m}^2$. To what temperature will it warm before it emits as much energy (from both sides) by radiation as it absorbs on the sunny side? Assume the sail is so thin that the temperature is uniform and no energy is emitted from the edges. Take the environment to be 0 K.

Solution

The rate at which solar energy arrives at the sail is given by $\mathcal{P}_{incident} = I \cdot A$, where I is the intensity of the solar radiation and A is the area exposed to the sunlight. This gives

$$\mathcal{P}_{incident} = \left(1.40 \times 10^3 \text{ W/m}^2\right)\left(1.00 \text{ km}^2\right)\left(\frac{10^3 \text{ m}}{1 \text{ km}}\right)^2 = 1.40 \times 10^9 \text{ W}$$

Of this incident radiation, 97.0% is reflected, and only 3.00% is absorbed. The rate at which the sail absorbs solar energy is then

$$\mathcal{P}_{absorbed} = 0.0300 \cdot \mathcal{P}_{incident} = 0.0300\left(1.40 \times 10^9 \text{ W}\right) = 4.20 \times 10^7 \text{ W}$$

Assuming the sail radiates equally from both sides, the total radiating area is $A' = 2A = 2(1.00 \text{ km})^2 = 2(1.00 \times 10^3 \text{ m})^2 = 2.00 \times 10^6 \text{ m}^2$. Stefan's law gives the rate at which this sail will radiate energy to a 0 K environment, when the sail is at absolute temperature T, as

$$\mathcal{P}_{rad} = \sigma A' e\left(T^4 - 0\right) = \left(5.669\ 6 \times 10^{-8} \text{ W/m}^2 \cdot \text{K}^4\right)\left(2.00 \times 10^6 \text{ m}^2\right)(0.03)T^4$$

or

$$\mathcal{P}_{rad} = \left(3.40 \times 10^{-3} \text{ W/K}^4\right)T^4$$

When the sail reaches its equilibrium temperature, it radiates energy away at the same rate as it is absorbing energy, or $\mathcal{P}_{rad} = \mathcal{P}_{absorbed}$. The equilibrium temperature is then

$$T = \left[\frac{\mathcal{P}_{absorbed}}{3.40 \times 10^{-3} \text{ W/K}^4}\right]^{1/4} = \left[\frac{4.20 \times 10^7 \text{ W}}{3.40 \times 10^{-3} \text{ W/K}^4}\right]^{1/4} = 333 \text{ K} \qquad \Diamond$$

55. A 200-g block of copper at a temperature of 90°C is dropped into 400 g of water at 27°C. The water is contained in a 300-g glass container. What is the final temperature of the mixture?

Solution

Assuming that the mixture is thermally isolated from the environment, the total energy absorbed by the originally cooler parts of the mixture (water and glass) must equal the energy given up by the originally hotter material (copper). Hence, $Q_{water} + Q_{glass} = -Q_{Cu}$, or

$$\left(m_w c_w + m_g c_g\right)\left(T_f - T_{i,cold}\right) = m_{Cu} c_{Cu}\left[-\left(T_f - T_{i,Cu}\right)\right]$$

which yields

$$\left(m_w c_w + m_g c_g + m_{Cu} c_{Cu}\right)T_f = m_{Cu} c_{Cu} T_{i,Cu} + \left(m_w c_w + m_g c_g\right)T_{i,cold}$$

and

$$T_f = \frac{m_{Cu} c_{Cu} T_{i,Cu} + \left(m_w c_w + m_g c_g\right)T_{i,cold}}{m_w c_w + m_g c_g + m_{Cu} c_{Cu}}$$

With

$$T_{i,Cu} = 90°C, \; T_{i,cold} = 27°C, \; m_{Cu} = 0.200 \text{ kg}, \; m_w = 0.400 \text{ kg}, \; m_g = 0.300 \text{ kg}$$

$$c_{Cu} = 387, \; c_g = 837 \text{ J/kg} \cdot °C, \; \text{and} \; c_w = 4\,186 \text{ J/kg} \cdot °C$$

this gives

$$T_f = \frac{(0.200)(387)(90) + \left[(0.400)(4\,186) + (0.300)(837)\right](27)}{(0.400)(4\,186) + (0.300)(837) + (0.200)(387)} = 29°C \qquad \Diamond$$

59. Water is being boiled in an open kettle that has a 0.500-cm-thick circular aluminum bottom with a radius of 12.0 cm. If the water boils away at a rate of 0.500 kg/min, what is the temperature of the lower surface of the bottom of the kettle? Assume the top surface of the bottom of the kettle is at 100°C.

Solution

The energy that must be added to evaporate a mass m of water is $Q = mL_v$. If water evaporates at a rate $\Delta m / \Delta t$, the required rate of energy input is

$$P = \frac{\Delta Q}{\Delta t} = \left(\frac{\Delta m}{\Delta t}\right)L_v = \left[\left(0.500 \; \frac{\text{kg}}{\text{min}}\right)\left(\frac{1 \text{ min}}{60.0 \text{ s}}\right)\right]\left(2.26 \times 10^6 \text{ J/kg}\right) = 1.88 \times 10^4 \text{ J/s}$$

Assuming that a steady state exists inside the kettle, this must also equal the rate at which energy is transferred through the aluminum bottom of the kettle by conduction. Thus,

$$P = kA\left(\frac{T_h - T_c}{L}\right) = 1.88 \times 10^4 \text{ W}$$

The area of the kettle bottom is $A = \pi r^2 = \pi (0.120 \text{ m})^2$, the thickness of the bottom is $L = 0.500$ cm $= 5.00 \times 10^{-3}$ m, the thermal conductivity of aluminum is $k = 238$ J/s·m·°C, and we know that the temperature of the upper surface of the kettle bottom is $T_c = 100$°C. Thus, the temperature of the lower surface of the kettle bottom must be

$$T_h = T_c + \frac{(1.88 \times 10^4 \text{ W})L}{kA} = 100°C + \frac{(1.88 \times 10^4 \text{ W})(5.00 \times 10^{-3} \text{ m})}{(238 \text{ J/s·m·°C})\pi (0.120 \text{ m})^2} = 109°C \qquad \lozenge$$

64. Three liquids are at temperatures of 10°C, 20°C, and 30°C, respectively. Equal masses of the first two liquids are mixed, and the equilibrium temperature is 17°C. Equal masses of the second and third are then mixed, and the equilibrium temperature is 28°C. Find the equilibrium temperature when equal masses of the first and third are mixed.

Solution

We are given that the masses of the liquid samples are $m_1 = m_2 = m_3$, and that the initial temperatures of the samples are $T_{i,1} = 10$°C, $T_{i,2} = 20$°C, and $T_{i,3} = 30$°C.

We assume that all of the mixtures are isolated from the environment, so $Q_{cold} = -Q_{hot}$ in each case.

When liquid 1 and liquid 2 are mixed, $T_f = 17$°C, giving

$$m_1 c_1 (17°C - 10°C) = -m_2 c_2 (17°C - 20°C) \qquad \text{or} \qquad c_1 = \frac{3}{7} c_2 \qquad \textbf{(1)}$$

When liquid 2 and liquid 3 are mixed, $T_f = 28$°C, and we have

$$m_2 c_2 (28°C - 20°C) = -m_3 c_3 (28°C - 30°C) \qquad \text{or} \qquad c_3 = \frac{8}{2} c_2 = 4 c_2 \qquad \textbf{(2)}$$

Then, when liquids 1 and 3 are mixed, we would find

$$m_1 c_1 (T_f - 10°C) = -m_3 c_3 (T_f - 30°C)$$

or

$$T_f = \frac{c_1 (10°C) + c_3 (30°C)}{c_1 + c_3}$$

Substituting from equations (1) and (2) yields

$$T_f = \frac{(3/7) c_2 (10°C) + (4) c_2 (30°C)}{(3/7) c_2 + (4) c_2} = \frac{3(10°C) + 28(30° C)}{3 + 28} = 28°C \qquad \lozenge$$

71. At time $t = 0$, a vessel contains a mixture of 10 kg of water and an unknown mass of ice in equilibrium at 0°C. The temperature of the mixture is measured over a period of an hour, with the following results: During the first 50 min, the mixture remains at 0°C; from 50 min to 60 min, the temperature increases steadily from 0°C to 2°C. Neglecting the heat capacity of the vessel, determine the mass of ice that was initially placed in it. Assume a constant power input to the container.

Solution

When thermal energy is added to a mixture of ice and water, all at the freezing point of water, the added energy initially goes into melting ice, and the temperature of the mixture will not begin to rise until all of the ice has been converted to liquid water. Once the ice is melted, the temperature of the melted ice and the original water starts to rise as energy continues to be added.

Let m be the mass of ice in the original mixture and P be the constant rate of adding thermal energy (or the power input). The thermal energy that must be added to melt all of the ice is $Q_1 = mL_f = m(3.33 \times 10^5 \text{ J/kg})$, where L_f is the latent heat of fusion of water. This quantity of energy equals the entire energy input during the first 50 minutes, or

$$Q_1 = m(3.33 \times 10^5 \text{ J/kg}) = P(50 \text{ min})(60 \text{ s/min}) = P(3.0 \times 10^3 \text{ s})$$

so the constant power input may be expressed as

$$P = \frac{Q_1}{3.0 \times 10^3 \text{ s}} = \frac{m(3.33 \times 10^5 \text{ J/kg})}{3.0 \times 10^3 \text{ s}}$$

or

$$P = m(1.11 \times 10^2 \text{ J/s} \cdot \text{kg})$$

The energy input in the last 10 minutes produces an increase in temperature of $\Delta T = 2.0°C$ in the $m_{total} = m + 10$ kg of liquid water, so we may write

$$Q_2 = m_{total} c_w (\Delta T) = (m + 10 \text{ kg})(4\,186 \text{ J/kg} \cdot °C)(2.0°C) = P(10 \text{ min})(60 \text{ s/min})$$

or

$$P = (m + 10 \text{ kg})\left(\frac{8.4 \times 10^3 \text{ J/kg}}{6.0 \times 10^2 \text{ s}}\right)$$

and

$$P = (m + 10 \text{ kg})(14 \text{ J/s} \cdot \text{kg})$$

Equating the two expressions we now have for the constant power input gives

$$m(1.11 \times 10^2 \text{ J/s} \cdot \text{kg}) = (m + 10 \text{ kg})(14 \text{ J/s} \cdot \text{kg})$$

or

$$m = \frac{(10 \text{ kg})(14 \text{ J/s} \cdot \text{kg})}{(1.11 \times 10^2 \text{ J/s} \cdot \text{kg} - 14 \text{ J/s} \cdot \text{kg})} = 1.4 \text{ kg}$$

12

The Laws of Thermodynamics

NOTES FROM SELECTED CHAPTER SECTIONS

12.1 Work in Thermodynamic Processes

The work done on a gas in a process that takes it from some initial state to some final state is the negative of the area under the curve on a *PV* diagram. (See the figure to the right.)

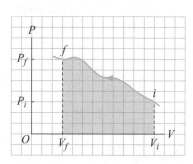

If the gas is compressed ($V_f < V_i$) the work done on the gas is positive.

If the gas expands ($V_f > V_i$) the work done on the gas is negative.

If the gas expands at constant pressure, called an **isobaric process,** then $W = -P(V_f - V_i)$.

$$\begin{pmatrix} \text{Work} \\ \text{on gas} \end{pmatrix} = - \begin{pmatrix} \text{Area under} \\ \text{the curve} \end{pmatrix}$$

The work done on a system depends on the process (path) by which the system goes from the initial to the final state; the quantity of work depends on the initial, final, and intermediate states of the system.

12.2 The First Law of Thermodynamics

In the **first law of thermodynamics,** $\Delta U = Q + W$, Q is the energy transferred to the system by heat and W is the work done on the system. Note that by convention, Q is positive when energy enters the system and negative when energy is removed from the system. Likewise, W can be positive or negative as mentioned earlier. The initial and final states must be equilibrium states; however, the intermediate states are, in general, non-equilibrium states since the thermodynamic coordinates undergo finite changes during the thermodynamic process.

Four basic thermodynamic processes as they apply to an ideal gas will be described in this chapter:

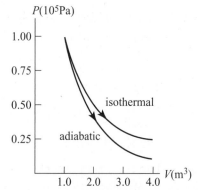

Isobaric (constant pressure)

Isovolumetric (constant volume)

Isothermal (constant temperature)

Adiabatic (no heat transfer)

Compare the *PV* diagrams for isothermal and adiabatic expansions illustrated in the figure at right. You should be able to add curves to the *PV* diagram that correspond to isobaric and isovolumetric processes.

12.3 Heat Engines and the Second Law of Thermodynamics

A **heat engine** is a device that converts internal energy to other useful forms, such as electrical and mechanical energy. A heat engine carries some working substance through a cyclic process during which (1) energy is transferred from a reservoir at a high temperature, (2) work is done by the engine, and (3) energy is expelled by the engine to a reservoir at a lower temperature.

The engine absorbs a quantity of energy, Q_h, from a hot reservoir, does work W_{eng} and then gives up energy Q_c to a cold reservoir. Because the working substance goes through a cycle, its initial and final internal energies are equal, so $\Delta U = 0$. Hence, from the first law of thermodynamics, *the work done by a heat engine equals the net energy absorbed from the reservoirs: $W_{eng} = Q_h - Q_c$.*

If the working substance is a gas, the work done by the engine for a cyclic process is the area enclosed by the curve representing the process on a PV diagram.

The **thermal efficiency, e,** of a heat engine is the ratio of the work done by the engine to the energy absorbed at the higher temperature during one cycle.

The second law of thermodynamics can be stated as follows:

> It is impossible to construct a heat engine that, operating in a cycle, produces no other effect than the absorption of heat from a reservoir and the performance of an equal amount of work.

A process is **irreversible** if the system and its surroundings cannot be returned to their initial states. A process is **reversible** if the system passes from the initial to the final state through a succession of equilibrium states and can be returned to its initial condition along the same path.

The **Carnot cycle** is the most efficient cyclic process (or engine) operating between two given energy reservoirs with temperatures T_h and T_c. The PV diagram for the Carnot cycle, illustrated in the figure, includes **four reversible processes: two isothermal and two adiabatic.**

The Carnot Cycle

1. The process $A \rightarrow B$ is an isotherm (constant T). During this process the gas expands at constant temperature T_h and absorbs energy Q_h from the hot reservoir.

2. The process $B \rightarrow C$ is an adiabatic expansion ($Q = 0$); the gas expands and cools to a temperature T_c.

3. The process $C \rightarrow D$ is a second isotherm (constant T); the gas is compressed at constant temperature T_c, and expels energy Q_c to the cold reservoir.

4. The final process $D \rightarrow A$ is an adiabatic compression ($Q = 0$) in which the gas temperature increases to a final temperature of T_h.

In practice, no working engine is 100% efficient, even when losses such as friction are neglected. One can obtain some theoretical limits on the efficiency of a real engine by comparison with the ideal Carnot engine. A **reversible engine** is one which will operate with the same efficiency in the forward and reverse directions. The Carnot engine is one example of a reversible engine.

> All Carnot engines operating reversibly between T_h and T_c have the same efficiency given by Equation (12.16).

> No real (irreversible) engine can have an efficiency greater than that of a reversible engine operating between the same two temperatures.

12.4 Entropy

Entropy is a quantity used to measure the **degree of disorder** in a system. For example, the molecules of a gas in a container at a high temperature are in a more disordered state (higher entropy) than the same molecules at a lower temperature.

When energy is transferred to a system by heat, the entropy increases. When energy is transferred out of a system by heat, the entropy decreases. In describing a thermodynamic process, the change in entropy is an important quantity; the entropy concept is quite powerful when analyzing a system that undergoes a change in state.

The second law of thermodynamics can be stated in terms of entropy as follows:

> *The total entropy of an isolated system always increases in time if the system undergoes an irreversible process. If an isolated system undergoes a reversible process, the total entropy remains constant.*

EQUATIONS AND CONCEPTS

Work done on a gas as the volume changes is equal to the negative of the area under the pressure vs. volume curve. When ΔV is negative (compression), the work done on the gas is positive; when ΔV is positive (expansion), the work done on the gas is negative. *Equation (12.1) applies if the pressure of the gas remains constant during a compression or an expansion.*

$$W = -P\Delta V \qquad (12.1)$$

Gas compressed at constant pressure.

The **first law of thermodynamics** is stated in mathematical form in Equation (12.2). The change in the internal energy of a system is equal to the sum of the energy transferred across the boundary by heat and the energy transferred by work. Q is positive when energy is added to the system by heat. W is positive when work is done on the system by its surroundings. *The values of both Q and W depend on the path or sequence of processes by which a system changes from an initial to a final state. The internal energy U depends only on the initial and final states of the system.*

$$\Delta U = Q + W \qquad (12.2)$$

Q = energy added to system by heat

W = work done on system

ΔU = change in the internal energy of system

Applications of the laws of thermodynamics will be described using some of the terms and equations described and shown below.

An **Isolated system** does not interact with the surroundings. *The internal energy of an isolated system remains constant.*

$$Q = W = 0$$
$$\Delta U = 0$$

In a **cyclic process** the initial and final states are the same and there is no change in the internal energy of the system. *The net work done **on** the gas per cycle equals the **negative** of the area enclosed by the path representing the process on a PV diagram.*

$$\Delta U = 0$$
$$Q = -W$$

In an **adiabatic process** no energy is transferred by heat ($Q = 0$) and the change in internal energy equals the work done on the system. *A system can undergo an adiabatic process if it is thermally insulated from its surroundings.*

$$\Delta U = W$$
$$PV^{\gamma} = \text{constant} \qquad (12.8a)$$

$$\gamma = \frac{C_p}{C_v} \qquad (12.8b)$$

C_p = molar heat capacity at constant pressure.

C_v = molar heat capacity at constant volume.

An **isobaric process** occurs at constant pressure and energy (Q) is transferred into (out of) the gas when it expands (contracts).

$$W = -P(V_f - V_i)$$
$$Q = nC_p \Delta T \qquad (12.6)$$

An **isovolumetric process** occurs at constant volume ($\Delta V = 0$) and zero work is done. *The net energy added to the system by heat at constant volume goes into increasing the internal energy.*

$$\Delta U = Q$$
$$Q = nC_v \Delta T \qquad (12.9)$$

An **isothermal process** occurs at constant temperature. *There is no change in internal energy ($\Delta U = 0$), and the work done on the system is equal to the negative of the energy added by heat.*

$$W = -Q$$
$$W_{env} = nRT \ln\left(\frac{V_f}{V_i}\right) \qquad (12.10)$$

W_{env} = work done on the environment

A **heat engine** is a device that converts thermal energy into other forms of energy by carrying a working substance through a cyclic process. During each cycle, the following occurs:

(1) Energy is transferred by heat from a high temperature reservoir.

(2) Work is accomplished by the engine.

(3) Energy is expelled to a low temperature source.

The net work (W_{eng}) done by a heat engine equals the net energy absorbed by the engine.

$$W_{eng} = |Q_h| - |Q_c| \qquad (12.11)$$

Q_h is the quantity of energy absorbed from the high temperature reservoir.

Q_c is the quantity of energy expelled to the low temperature reservoir.

The **thermal efficiency** of a heat engine is the ratio of the work done to the energy absorbed at the higher temperature during one cycle of the process.

$$e \equiv \frac{W_{eng}}{|Q_h|} = \frac{|Q_h| - |Q_c|}{|Q_h|} = 1 - \frac{|Q_c|}{|Q_h|} \qquad (12.12)$$

Carnot efficiency is the maximum theoretical efficiency of a heat engine operating in an ideal reversible cycle between two reservoirs of temperatures T_h and T_c. *All real engines are less efficient than the Carnot engine.*

$$e_C = 1 - \frac{T_c}{T_h} \qquad (12.16)$$

The **coefficient of performance (COP)** is a measure of the effectiveness of a refrigerator or heat pump.

$$\text{COP(cooling mode)} = \frac{|Q_c|}{W} \qquad (12.13)$$

$$\text{COP(heating mode)} = \frac{|Q_h|}{W} \qquad (12.14)$$

Entropy, *S,* is a thermodynamic variable which characterizes the degree of disorder in a system. *All physical processes tend toward a state of increasing entropy.*

The **change in entropy** as a system goes from an initial to a final state, is the ratio of the energy transferred to the system along a reversible path to the absolute temperature of the system.

$$\Delta S \equiv \frac{\Delta Q_r}{T} \qquad (12.17)$$

REVIEW CHECKLIST

- Understand how work is defined when a system undergoes a change in state, and the fact that work depends on the path taken by the system. You should also know how to sketch processes on a *PV* diagram, and calculate work using these diagrams.

- State the first law of thermodynamics ($\Delta U = Q + W$), and explain the meaning of the three forms of energy contained in this statement. Discuss the implications of the first law of thermodynamics as applied to (i) an isolated system, (ii) a cyclic process, (iii) an adiabatic process, and (iv) an isothermal process.

- Describe the processes via which an ideal heat engine goes through a **Carnot cycle.** Express the efficiency of an ideal heat engine (Carnot engine) as a function of work and energy exchange with its environment. Express the maximum efficiency of an ideal heat engine as a function of its input and output temperatures.

- Understand the concept of entropy. Define change in entropy for a system in terms of its energy gain or loss by heat, and its temperature. State the second law of thermodynamics as it applies to entropy changes in a thermodynamic system.

SOLUTIONS TO SELECTED END-OF-CHAPTER PROBLEMS

3. Gas in a container is at a pressure of 1.5 atm and a volume of 4.0 m³. What is the work done *on* the gas (a) if it expands at constant pressure to twice its initial volume, and (b) if it is compressed at constant pressure to one-quarter its initial volume?

Solution

(a) In an isobaric process, the work done *on* the system is

$$W = -P(\Delta V) = -P\left(V_f - V_i\right)$$

Thus, if the constant pressure is $P = 1.5$ atm and $V_i = 4.0$ m³, the work done *on* the gas as it expands to twice its initial volume is

$$W = -(1.5 \text{ atm})\left(2V_i - V_i\right) = -1.5\left(1.013 \times 10^5 \text{ Pa}\right)\left(4.0 \text{ m}^3\right) = -6.1 \times 10^5 \text{ J} \qquad \Diamond$$

(b) If the gas is compressed to one-quarter of its initial volume, the work done *on* the gas is

$$W = -P(\Delta V) = -P\left(V_f - V_i\right) = -P\left(\frac{V_i}{4} - V_i\right) = +\frac{3}{4}PV_i$$

or

$$W = +\frac{3}{4}\left[1.5\left(1.013 \times 10^5 \text{ Pa}\right)\right]\left(4.0 \text{ m}^3\right) = +4.6 \times 10^5 \text{ J} \qquad \Diamond$$

7. A sample of helium behaves as an ideal gas as it is heated at constant pressure from 273 K to 373 K. If 20.0 J of work is done by the gas during this process, what is the mass of helium present?

Solution

When pressure is constant, the ideal gas law gives $PV_f - PV_i = nRT_f - nRT_i$, or $P(\Delta V) = nR(\Delta T)$. But, in an isobaric process, the work a system does on its environment is $W_{env} = P(\Delta V)$, so

$$W_{env} = nR(\Delta T) \qquad (1)$$

where $n = m/M$ is the number of moles present, m is the mass of the gas, and M is the molecular weight. For helium, $M_{He} = 4.00$ g/mol, and if the work done *by* the gas is $W = 20.0$ J, we have

$$W_{env} = \left(\frac{m}{M_{He}}\right)R(\Delta T)$$

or

$$m = \frac{W_{env}M_{He}}{R(\Delta T)}$$

which yields

$$m = \frac{(20.0 \text{ J})(4.00 \text{ g/mol})}{(8.31 \text{ J/mol}\cdot\text{K})(373 \text{ K} - 273 \text{ K})} = 9.63 \times 10^{-2} \text{ g} = 0.963 \text{ mg} \qquad \Diamond$$

As an alternate approach to this problem, we could make use of the properties of an ideal gas and the first law of thermodynamics. For an ideal gas, the energy added by heat is $Q = nC_p(\Delta T)$, and the change in the internal energy is $\Delta U = nC_v(\Delta T)$. Here, C_v is the molar heat capacity at constant volume, and $C_p = C_v + R$ is the molar heat capacity at constant pressure for the ideal gas. For a monatomic ideal gas, such as helium, $C_v = \frac{3}{2}R$ and $C_p = \frac{5}{2}R$.

The first law of thermodynamics is $\Delta U = Q + W = Q - W_{env}$, where $W = -W_{env}$ is the work done *on* the system. Thus, for this monatomic, ideal gas, we have

$$W_{env} = Q - \Delta U = n\left(\tfrac{5}{2}R\right)\Delta T - n\left(\tfrac{3}{2}R\right)\Delta T$$

or

$$W_{env} = nR(\Delta T)$$

which is the same as equation (1) above. Thus, the two methods of solution are fully equivalent and will yield the same results.

---

13. A gas expands from I to F in Figure P12.5. The energy added to the gas by heat is 418 J when the gas goes from I to F along the diagonal path. (a) What is the change in internal energy of the gas? (b) How much energy must be added to the gas by heat for the indirect path IAF to give the same change in internal energy?

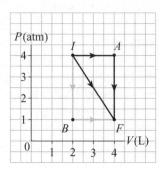

Figure P12.5

Solution

(a) The area under the diagonal path IF is a triangle, of height 3.00 atm, sitting atop a rectangle with a height of 1.00 atm. The base of the triangle is the same as the base of the rectangle, 2.00 L. The work done on the gas during this process is then

$$W = -(\text{area under curve}) = -\frac{1}{2}(3.00 \text{ atm})(2.00 \text{ L}) - (1.00 \text{ atm})(2.00 \text{ L})$$

or

$$W = -5.00 \text{ atm} \cdot \text{L}\left(\frac{1.013 \times 10^5 \text{ N/m}^2}{1 \text{ atm}}\right)\left(\frac{10^{-3} \text{ m}^3}{1 \text{ L}}\right) = -506.5 \text{ J}$$

Since the energy added to the gas by heat is $Q = 418$ J for this process, the first law of thermodynamics gives the change in the internal energy as

$$U_F - U_I = \Delta U = Q + W = 418 \text{ J} - 506.5 \text{ J} = -88.5 \text{ J} \qquad \lozenge$$

(b) The area under the indirect path *IAF* is a simple rectangle, of height 4.00 atm and width 2.00 L. The work done on the gas during this process is

$$W = -(\text{area under curve}) = -(4.00 \text{ atm})(2.00 \text{ L}) = -8.00 \text{ atm} \cdot \text{L}$$

or

$$W = -8.00 \text{ atm} \cdot \text{L} \left(\frac{1.013 \times 10^5 \text{ N/m}^2}{1 \text{ atm}} \right) \left(\frac{10^{-3} \text{ m}^3}{1 \text{ L}} \right) = -810 \text{ J}$$

The change in the internal energy of the gas between states *F* and *I* is the same as found above, $\Delta U = U_F - U_I = -88.5$ J. The first law of thermodynamics, $\Delta U = Q + W$, then gives the energy added to the gas by heat in the process *IAF* as

$$Q = \Delta U - W = -88.5 \text{ J} - (-810 \text{ J}) = 722 \text{ J} \qquad \Diamond$$

17. A gas is enclosed in a container fitted with a piston of cross-sectional area 0.150 m². The pressure of the gas is maintained at 6 000 Pa as the piston moves inward 20.0 cm. (a) Calculate the work done by the gas. (b) If the internal energy of the gas decreases by 8.00 J, find the amount of heat removed from the system by heat during the compression.

Solution

(a) In an isobaric compression such as this, the work done *on* the system is $W = -P(\Delta V)$. For this piston and cylinder, the change in volume of the trapped gas is $\Delta V = -A \cdot \Delta x$ where Δx is the distance the piston moves inward. Thus,

$$W = -(6.00 \times 10^3 \text{ Pa})[-(0.150 \text{ m}^3)(0.200 \text{ m})] = +180 \text{ J}$$

The work done *by* the gas is then

$$W_{\text{env}} = -W = -(180 \text{ J}) = -180 \text{ J} \qquad \Diamond$$

(b) The first law of thermodynamics says that the change in the internal energy of a system, ΔU, equals the sum of the energy transferred to the system by heat, Q, and the work done *on* the system, W. Thus, for the piston-cylinder-gas system of this problem, $\Delta U = Q + W$, or the energy added by heat is

$$Q = \Delta U - W = -8.00 \text{ J} - (+180 \text{ J}) = -188 \text{ J}$$

The energy transferred from the system to the environment (i.e., removed from the system) by heat is then

$$Q_{\text{env}} = -Q = -(-188 \text{ J}) = +188 \text{ J} \qquad \Diamond$$

25. A 5.0-kg block of aluminum is heated from 20°C to 90°C at atmospheric pressure. Find (a) the work done by the aluminum, (b) the amount of energy transferred to it by heat, and (c) the increase in its internal energy.

Solution

(a) Using the density from Table 9.3, the initial volume of the aluminum block is

$$V_0 = \frac{m}{\rho_{Al}} = \frac{5.0 \text{ kg}}{2.70 \times 10^3 \text{ kg/m}^3} = 1.85 \times 10^{-3} \text{ m}^3$$

As the block is heated, it undergoes an increase in volume given by

$$\Delta V = \beta_{Al} V_0 \Delta T = (3\alpha_{Al}) V_0 \Delta T$$
$$= 3 \left[24 \times 10^{-6} \text{ (°C)}^{-1} \right] (1.85 \times 10^{-3} \text{ m}^3)(90°C - 20°C) = 9.3 \times 10^{-6} \text{ m}^3$$

where the coefficient of linear expansion was obtained from Table 10.1. The work done *by* the aluminum is then

$$W_{\text{by block}} = +P(\Delta V) = (1.013 \times 10^5 \text{ Pa})(9.3 \times 10^{-6} \text{ m}^3) = 0.95 \text{ J} \qquad \Diamond$$

(b) The energy transferred to the aluminum by heat during this process is found, using the specific heat of aluminum from Table 11.1, as

$$Q = mc_{Al}(\Delta T) = (5.0 \text{ kg})(900 \text{ J/kg} \cdot °C)(90°C - 20°C) = 3.2 \times 10^5 \text{ J} \qquad \Diamond$$

(c) From the first law of thermodynamics, the change in the internal energy of the block is given by $\Delta U = Q + W$ where W is the work done *on* the aluminum block. Hence,

$$W = -W_{\text{by block}} = -0.95 \text{ J}$$

and

$$\Delta U = Q + W = 3.2 \times 10^5 \text{ J} - 0.95 \text{ J} = 3.2 \times 10^5 \text{ J} \qquad \Diamond$$

35. One of the most efficient engines ever built is a coal-fired steam turbine engine in the Ohio River valley, driving an electric generator as it operates between 1 870°C and 430°C. (a) What is its maximum theoretical efficiency? (b) Its actual efficiency is 42.0%. How much mechanical power does the engine deliver if it absorbs 1.40×10^5 J of energy each second from the hot reservoir?

Solution

(a) The maximum theoretical efficiency is that of a Carnot engine operating between the specified reservoirs. The absolute temperatures of the given hot and cold reservoirs are

$$T_h = 1\,870 + 273 = 2\,143 \text{ K} \qquad \text{and} \qquad T_c = 430 + 273 = 703 \text{ K}$$

The Carnot efficiency is then

$$e_c = \frac{T_h - T_c}{T_h} = 1 - \frac{T_c}{T_h} = 1 - \frac{703 \text{ K}}{2\,143 \text{ K}} = 0.672 \quad (\text{or } 67.2\%) \qquad \diamond$$

(b) The actual efficiency of a heat engine is defined as $e = W_{eng} / |Q_h|$, where W_{eng} is the work done by the engine during some interval and $|Q_h|$ is the energy absorbed from the hot reservoir during the same interval.

This engine absorbs $|Q_h| = 1.40 \times 10^5$ J from the hot reservoir each second and has an actual efficiency of $e = 0.420$ (42.0%). Therefore, the work done by the engine each second (or the power delivered) is

$$\mathcal{P} = \frac{W_{eng}}{\Delta t} = \frac{e|Q_h|}{\Delta t} = \frac{(0.420)(1.40 \times 10^5 \text{ J})}{1.00 \text{ s}} = 5.88 \times 10^4 \text{ W} = 58.8 \text{ kW} \qquad \diamond$$

41. In one cycle a heat engine absorbs 500 J from a high temperature reservoir and expels 300 J to a low temperature reservoir. If the efficiency of this engine is 60% of the efficiency of a Carnot engine, what is the ratio of the low temperature to the high temperature in the Carnot engine?

Solution

The efficiency of a heat engine is $e = W_{eng}/|Q_h|$, where W_{eng} is the work done by the engine, and $|Q_h|$ is the energy drawn from the hot (or higher temperature) reservoir. But, $W_{eng} = |Q_h| - |Q_c|$, where $|Q_c|$ is the energy exhausted to the cold (or lower temperature) reservoir. Thus, the efficiency may be written as

$$e = \frac{W_{eng}}{|Q_h|} = \frac{|Q_h| - |Q_c|}{|Q_h|} = 1 - \frac{|Q_c|}{|Q_h|} = 1 - \frac{300 \text{ J}}{500 \text{ J}} = 0.40 \text{ (or 40\%)}$$

If the efficiency of our actual machine is 60% of the Carnot efficiency, then $e = 0.60 \cdot e_C$, or the efficiency of a Carnot engine operating between the thermal energy reservoirs used by our engine would be

$$e_C = \frac{e}{0.60} = \frac{0.40}{0.60} = \frac{2}{3}$$

Since the efficiency of a Carnot engine is expressed in terms of the absolute temperatures of its reservoirs as

$$e_C = \frac{T_h - T_c}{T_h} = 1 - \frac{T_c}{T_h}$$

the ratio of the temperature of the low temperature reservoir to that of the high temperature reservoir must be

$$\frac{T_c}{T_h} = 1 - e_C = 1 - \frac{2}{3} = \frac{1}{3} \qquad \qquad \Diamond$$

47. A freezer is used to freeze 1.0 L of water completely into ice. The water and the freezer remain at a constant temperature of $T = 0°C$. Determine (a) the change in the entropy of the water and (b) the change in the entropy of the freezer.

Solution

The change in the entropy of a thermodynamic system during a reversible, constant temperature, process is given by

$$\Delta S = \frac{Q_r}{T}$$

where Q_r is the energy added to the system by heat and T is the absolute temperature of the system during this process.

As 1.0-L of liquid water at 0°C is converted to ice, the quantity of energy transferred by heat *from* the water and *to* the freezer is

$$|Q_r| = mL_f = (\rho_{water}V)L_f = \left[(10^3 \text{ kg/m}^3)(1.0 \text{ L})\left(\frac{10^{-3} \text{ m}^3}{1 \text{ L}}\right)\right](3.33\times10^5 \text{ J/kg})$$

or

$$|Q_r| = 3.3\times10^5 \text{ J} = 330 \text{ kJ}$$

The constant temperature of both the water and freezer during this process is

$$T = 0°C = 273 \text{ K}$$

(a) Energy is transferred from the water during the freezing process, so $(Q_r)_{water} < 0$ and the change in the entropy of the water is

$$\Delta S_{water} = \frac{(Q_r)_{water}}{T} = \frac{-330 \text{ kJ}}{273 \text{ K}} = -1.2 \text{ kJ/K} \qquad \Diamond$$

(b) Energy is transferred to the freezer by heat, so $(Q_r)_{freezer} > 0$ and

$$\Delta S_{freezer} = \frac{(Q_r)_{freezer}}{T} = \frac{+330 \text{ kJ}}{273 \text{ K}} = +1.2 \text{ kJ/K} \qquad \Diamond$$

Note that in this **reversible** process, the total entropy of the isolated system consisting of (freezer plus water) is constant, or

$$\Delta S_{\substack{isolated \\ system}} = \Delta S_{water} + \Delta S_{freezer} = 0$$

51. The surface of the Sun is approximately at 5 700 K, and the temperature of Earth's surface is approximately 290 K. What entropy change occurs when 1 000 J of energy is transferred by heat from the Sun to Earth?

Solution

A quantity of energy $|Q|$ is transferred *from* the Sun at absolute temperature T_{Sun}. The change in the entropy of the Sun during this process is

$$\Delta S_{Sun} = \frac{Q_r}{T_{Sun}} = \frac{-|Q|}{T_{Sun}}$$

This same quantity of energy is transferred *to* Earth at absolute temperature T_{Earth}, and the change in the entropy of Earth for this process is

$$\Delta S_{Earth} = \frac{Q_r}{T_{Earth}} = \frac{+|Q|}{T_{Earth}}$$

The total change in entropy of the Universe during this energy exchange is then

$$\Delta S_{total} = \Delta S_{Sun} + \Delta S_{Earth} = \frac{-|Q|}{T_{Sun}} + \frac{+|Q|}{T_{Earth}} = |Q|\left(\frac{1}{T_{Earth}} - \frac{1}{T_{Sun}}\right)$$

Thus, if $|Q| = 1\ 000$ J, $T_{Sun} = 5\ 700$ K, and $T_{Earth} = 290$ K, the total change in entropy is

$$\Delta S_{total} = (1\ 000\ \text{J})\left(\frac{1}{290\ \text{K}} - \frac{1}{5\ 700\ \text{K}}\right) = +3.27 \ \text{J/K} \qquad \Diamond$$

Note that we have treated both the withdrawal of energy from the Sun and the addition of energy to Earth as constant temperature processes. The justification for this is that the energy transferred $(|Q| = 1\ 000\ \text{J})$ is very small in comparison to the total internal energy of each reservoir (Sun and Earth) involved in the transfer.

59. A 1 500-kW heat engine operates at 25% efficiency. The heat energy expelled at the low temperature is absorbed by a stream of water that enters the cooling coils at 20°C. If 60 L flows across the coils per second, determine the increase in temperature of the water.

Solution

The power output from the engine is

$$P = \frac{W_{eng}}{\Delta t} = 1\,500 \text{ kW} = 1\,500 \times 10^3 \text{ W} = 1.5 \times 10^6 \text{ J/s}$$

so the work done by this engine in a time interval of $\Delta t = 1.0$ s is $W_{eng} = 1.5 \times 10^6$ J.

The efficiency of a heat engine is defined as $e = W_{eng}/|Q_h|$. If the efficiency of this engine is $e = 0.25$ (25%), the energy absorbed by heat from the high temperature (or hot) reservoir each second is

$$|Q_h| = \frac{W_{eng}}{e} = \frac{1.5 \times 10^6 \text{ J}}{0.25} = 6.0 \times 10^6 \text{ J}$$

and the energy exhausted by heat to the low temperature (or cold) reservoir each second is

$$|Q_c| = |Q_h| - W_{eng} = 6.0 \times 10^6 \text{ J} - 1.5 \times 10^6 \text{ J} = 4.5 \times 10^6 \text{ J}$$

The mass of the 60 L of 20°C water that flows over the cooling coils during this 1 second time interval is

$$m = \rho_{water} V = \left(1.0 \times 10^3 \ \frac{\text{kg}}{\text{m}^3}\right)(60 \ \text{L})\left(\frac{10^{-3} \ \text{m}^3}{1 \ \text{L}}\right) = 60 \text{ kg}$$

and the increase in the temperature of the cooling water will be

$$\Delta T = \frac{|Q_c|}{mc_{water}} = \frac{4.5 \times 10^6 \text{ J}}{(60 \text{ kg})(4\,186 \text{ J/kg} \cdot {}^\circ\text{C})} = 18{}^\circ\text{C} \qquad \Diamond$$

67. A cylinder containing 10.0 moles of a monatomic ideal gas expands from Ⓐ to Ⓑ along the path shown in Figure P12.67. (a) Find the temperature of the gas at point Ⓐ and the temperature at point Ⓑ. (b) How much work is done by the gas during this expansion? (c) What is the change in internal energy of the gas? (d) Find the energy transferred to the gas by heat in this process?

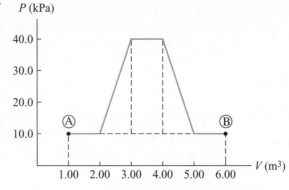

Figure P12.67

Solution

(a) At point A, the gas pressure is $P_A = 10.0$ kPa $= 10.0 \times 10^3$ Pa $= 1.00 \times 10^4$ Pa and its volume is $V_A = 1.00$ m^3. The ideal gas law then gives its temperature as

$$T_A = \frac{P_A V_A}{nR} = \frac{\left(1.00 \times 10^4 \text{ Pa}\right)\left(1.00 \text{ m}^3\right)}{10.0\left(8.31 \text{ J/kg}\cdot\text{K}\right)} = 1.20 \times 10^2 \text{ K} \qquad \lozenge$$

Similarly, at point B, $P_B = 1.00 \times 10^4$ Pa, $V_B = 6.00$ m^3, and

$$T_B = \frac{P_B V_B}{nR} = \frac{\left(1.00 \times 10^4 \text{ Pa}\right)\left(6.00 \text{ m}^3\right)}{10.0\left(8.31 \text{ J/kg}\cdot\text{K}\right)} = 722 \text{ K} \qquad \lozenge$$

(b) The work done *by* the gas (since it is expanding) on the environment equals the area under the process curve between points A and B. Observe from Figure 12.67 that this area can be computed as the sum of two rectangular areas and two triangular areas. This gives

$$W_{\text{env}} = \left(10 \times 10^3 \text{ Pa}\right)\left(6.00 \text{ m}^3 - 1.00 \text{ m}^3\right) + \left[\left(40.0 - 10.0\right) \times 10^3 \text{ Pa}\right]\left(4.00 - 3.00\right) \text{ m}^3$$

$$+ \frac{1}{2}\left[\left(40.0 - 10.0\right) \times 10^3 \text{ Pa}\right]\left(3.00 - 2.00\right) \text{ m}^3$$

$$+ \frac{1}{2}\left[\left(40.0 - 10.0\right) \times 10^3 \text{ Pa}\right]\left(5.00 - 4.00\right) \text{ m}^3$$

or

$$W_{\text{env}} = 1.10 \times 10^5 \text{ J} \qquad \lozenge$$

(c) Since this is a monatomic, ideal gas, the change in its internal energy during this process is

$$\Delta U = nC_v\Delta T = \frac{3}{2}nR(\Delta T) = \frac{3}{2}nR(T_B - T_A)$$

or

$$\Delta U = \frac{3}{2}(10.0 \text{ mol})(8.31 \text{ J/mol}\cdot\text{K})(722 \text{ K} - 120 \text{ K}) = 7.50\times10^4 \text{ J} \qquad \lozenge$$

(d) Using the first law of thermodynamics, the energy transferred to the gas by heat during this process is found to be

$$Q = \Delta U - W = \Delta U - (-W_{\text{env}}) = \Delta U + W_{\text{env}}$$

or

$$Q = 7.50\times10^4 \text{ J} + 1.10\times10^5 \text{ J} = 1.85\times10^5 \text{ J} \qquad \lozenge$$

---

71. An electrical power plant has an overall efficiency of 15%. The plant is to deliver 150 MW of electrical power to a city, and its turbines use coal as fuel. The burning coal produces steam at 190°C, which drives the turbines. The steam is condensed into water at 25°C by passing through coils that are in contact with river water. (a) How many metric tons of coal does the plant consume each day $(1 \text{ metric ton} = 1\times10^3 \text{ kg})$? (b) What is the total cost of the fuel per year if the delivery price is \$8 per metric ton? (c) If the river water is delivered at 20°C, at what minimum rate must it flow over the cooling coils so that its temperature does not exceed 25°C? (*Note:* The heat of combustion of coal is 7.8×10^3 cal/g.)

Solution

(a) If the plant has an overall efficiency of 15% and is to have a power output of 150 MW, the required power input (from combustion of coal) is

$$\mathcal{P}_{\text{input}} = \frac{\mathcal{P}_{\text{output}}}{e} = \frac{150\times10^6 \text{ W}}{0.15} = 1.0\times10^9 \text{ J/s}$$

The rate at which coal must be burned to yield this needed power input is

$$\frac{\Delta m}{\Delta t} = \frac{\mathcal{P}_{\text{input}}}{\text{heat of combustion}} = \frac{(1.0\times10^9 \text{ J/s})(8.64\times10^4 \text{ s/d})}{\left(7.8\times10^3 \dfrac{\text{cal}}{\text{g}}\right)\left(\dfrac{10^3\text{g}}{1\text{ kg}}\right)\left(\dfrac{4.186\text{ J}}{1\text{ cal}}\right)} = 2.64\times10^6 \text{ kg/d}$$

or

$$\frac{\Delta m}{\Delta t} = \left(2.64\times10^6 \frac{\text{kg}}{\text{d}}\right)\left(\frac{1 \text{ metric ton}}{10^3\text{ kg}}\right) \approx 2.6\times10^3 \text{ metric ton/d} \qquad \lozenge$$

(b) The annual fuel cost will be $cost = rate \times (\Delta m/\Delta t) \times (\Delta t)$

or

$$cost = \left(\frac{\$8}{\text{metric ton}} \right) \left(2.64 \times 10^3 \; \frac{\text{metric ton}}{\text{d}} \right) \left(\frac{365.242 \text{ d}}{1 \text{ y}} \right) = \$7.7 \times 10^6/\text{y} \qquad \Diamond$$

(c) The rate at which cooling waters must absorb exhaust energy is

$$\mathcal{P}_{\text{exhaust}} = \mathcal{P}_{\text{input}} - \mathcal{P}_{\text{output}} = 1.0 \times 10^9 \text{ W} - 150 \times 10^6 \text{ W} = 8.5 \times 10^8 \text{ W}$$

If the maximum rise in temperature of the cooling waters is to be $\Delta T = 5.0°\text{C}$, the minimum flow rate of the water is

$$flow \; rate = \frac{\Delta m_{\text{water}}}{\Delta t} = \frac{\mathcal{P}_{\text{exhaust}}}{c_{\text{water}}(\Delta T)} = \frac{8.5 \times 10^8 \text{ J/s}}{(4\,186 \text{ J/kg} \cdot °\text{C})(5.0°\text{C})} = 4.1 \times 10^4 \text{ kg/s} \qquad \Diamond$$

13

Vibrations and Waves

NOTES FROM SELECTED CHAPTER SECTIONS

13.1 Hooke's Law

13.2 Elastic Potential Energy

Simple harmonic motion occurs when the net force along the direction of motion is a Hooke's law type of force; that is, when the net force is proportional to the displacement and in the opposite direction.

It is necessary to define a few terms relative to harmonic motion:

- **The amplitude,** A, is the maximum distance that an object moves away from its equilibrium position. In the absence of friction, an object will continue in simple harmonic motion and reach a maximum distance equal to the amplitude on each side of the equilibrium position during each cycle.

- **The period,** T, is the time it takes the object to execute one complete cycle of the motion.

- **The frequency,** f, is the number of cycles or vibrations per unit of time.

Oscillatory motions are exhibited by many physical systems such as a mass attached to a spring, a pendulum, atoms in a solid, stringed musical instruments, and electrical circuits driven by a source of alternating current. *Simple harmonic motion of a mechanical system corresponds to the oscillation of an object between two points for an indefinite period of time, with no loss in mechanical energy.*

An object exhibits simple harmonic motion if the net external force acting on it is a linear restoring force.

201

13.4 Position, Velocity, and Acceleration as Functions of Time

The position (x), velocity (v), and acceleration (a) of an object moving with simple harmonic motion are shown in the three graphs below.

In this particular case, the object was released from rest when it was a maximum distance (amplitude) from the equilibrium position.

Shown here, from top to bottom, are graphs of displacement, velocity, and acceleration versus time for an object moving with simple harmonic motion under the initial conditions that $x = A$ and $v = 0$ at $t = 0$.

At the instant indicated by the dashed line:

$t = (3T/4)$

$x = 0$ (moving in the +x direction)

$v = +\omega A$ (maximum positive value)

$a = 0$

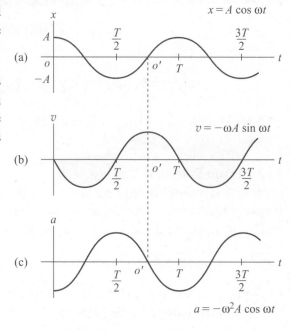

$$\omega = \left(\frac{2\pi}{T}\right) \quad \text{(in radians)}$$

A common system which undergoes simple harmonic motion is the **mass-spring system** shown in the figure at the right. The mass is assumed to move on a horizontal, frictionless surface. The point $x = 0$ is the equilibrium position of the mass; that is, the point where the mass would reside if left undisturbed. In this position, there is no horizontal force on the mass. When the mass is displaced a distance x from its equilibrium position, the spring produces a linear restoring force given by Hooke's law, $F = -kx$. The parameter k is the force constant of the spring and has SI units of N/m. The minus sign means that $\vec{\mathbf{F}}$ is directed toward the left when the displacement, x, is positive and is directed toward the right when x is negative. *The direction of the force $\vec{\mathbf{F}}$ is always toward the equilibrium position.*

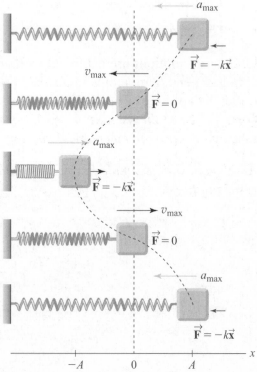

13.5 Motion of a Pendulum

A **simple pendulum** consists of a mass m attached to a light string of length L as shown in the figure. When the maximum angular displacement is small (θ less than approximately 15°), the pendulum exhibits simple harmonic motion. In this case, the resultant force acting on the mass m equals the component of weight tangent to the circular path followed by the mass. The magnitude of the resultant force equals $mg\sin\theta$. Since this force is always directed toward $\theta = 0$, it corresponds to a restoring force.

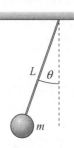

The period depends only on the length of the pendulum and the acceleration due to gravity. The period does not depend on mass, so we conclude that all simple pendula of equal length oscillate with the same frequency or period at the same location.

13.6 Damped Oscillations

Damped oscillations occur in realistic systems in which retarding forces such as friction are present. These forces will reduce the amplitudes of the oscillations with time, since mechanical energy is continually transferred from the oscillating system. Depending on the value of the frictional (retarding) force three distinct types of damping can be identified:

Underdamped—In this case, the retarding force is small compared to the restoring force. For small damping, oscillations continue with a frequency slightly less than that of the undamped system. The amplitude of the motion decreases exponentially with time until the oscillations cease at zero amplitude.

Critically damped—With an increase in the retarding force the system does not oscillate and returns to the equilibrium position in the shortest possible time without passing through the equilibrium point.

Overdamped—This mode is similar to critical damping and occurs with further increase in the frictional force. In this case the system also returns to equilibrium without passing through the equilibrium point but requires a longer time to do so.

It is possible to compensate for the energy lost in a damped oscillator by adding an additional driving force that does positive work on the system. This additional energy supplied to the system must equal the energy lost due to friction to maintain constant amplitude.

13.7 Waves

The production of mechanical waves requires (1) an **elastic medium** that can be disturbed, (2) **an energy source** to provide a disturbance or deformation in the medium, and (3) a **physical mechanism** by way of which adjacent portions of the medium can influence each other. The three parameters important in characterizing waves are (1) **wavelength,** (2) **frequency,** and (3) **wave velocity.**

Transverse waves, shown below in figure (a), are those in which particles of the disturbed medium move along a direction which is perpendicular to the direction of the wave velocity. For **longitudinal waves,** shown figure (b), the particles of the medium undergo displacements which are parallel to the direction of wave motion.

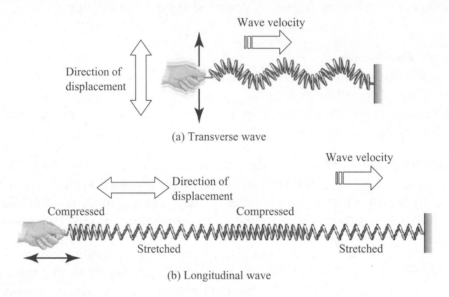

(a) Transverse wave

(b) Longitudinal wave

13.8 Frequency, Amplitude, and Wavelength

Consider a wave traveling in the x-direction on a very long string. Each particle along the string oscillates in simple harmonic motion along the y-direction with a **frequency** equal to the frequency of the source producing the vibration. The maximum distance the string is displaced above or below the

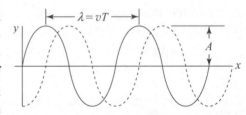

equilibrium value is called the **amplitude,** A, of the wave. The distance between two successive points along the string having maximum upper displacement is called the **wavelength,** λ. The number of crests or peaks of the wave passing a fixed point each second is the wave frequency (or harmonic frequency). The wave will advance along the string a distance of one wavelength in a time interval equal to one **period** of vibration, T; the time required for any point in the medium to complete one full cycle of vibration.

13.9 The Speed of Waves on Strings

For linear waves, the **speed of mechanical waves** depends only on the physical properties of the medium through which the disturbance travels. In the case of **waves on a string,** the velocity depends on the tension in the string and its mass per unit length (linear mass density).

13.10 Interference of Waves

If two or more waves are moving through a medium, the **resultant wave function** is the algebraic sum of the wave functions of the individual waves. Two traveling waves can pass through each other without being destroyed or altered.

13.11 Reflection of Waves

Whenever a traveling wave reaches a boundary, **reflection** occurs; part or all of the wave is reflected back through the original medium. If the wave is traveling along a string and is reflected from a "fixed" end, the reflected pulse is inverted. By contrast, a pulse is reflected without inversion at the "free" end of a string.

EQUATIONS AND CONCEPTS

Hooke's law gives the force exerted by a spring that is stretched or compressed beyond the equilibrium position. The force constant, k, is always positive and has a value which corresponds to the relative stiffness of the spring. *The negative sign in Equation (13.1) indicates that the force exerted by the string on the mass is always directed opposite the displacement toward the equilibrium position.*

$$F_s = -kx \qquad (13.1)$$

In simple harmonic motion, $\vec{\mathbf{F}} = -k\vec{\mathbf{x}}$.

Acceleration of a mass in simple harmonic motion has a magnitude proportional to the displacement and oppositely directed. *Note that the acceleration of a harmonic oscillator does not remain constant during the motion.* The acceleration equals zero when the oscillating mass is at the equilibrium position ($x = 0$) and has its maximum magnitude when $x = \pm A$ (the amplitude).

$$a = -\left(\frac{k}{m}\right)x \qquad (13.2)$$

Acceleration as a function of position

Elastic potential energy is stored in the spring as a result of work done by an external force to stretch or compress the spring.

$$PE_s \equiv \tfrac{1}{2}kx^2 \qquad (13.3)$$

Conservation of mechanical energy in a spring-mass system (including kinetic energy, gravitational potential energy, and elastic potential energy) holds when only conservative forces are acting on the system.

$$(KE + PE_g + PE_s)_i = (KE + PE_g + PE_s)_f$$

(13.4)

The **speed of an object in simple harmonic motion** is a maximum at $x = 0$; the speed is zero when the mass is at the points of maximum displacement $(x = \pm A)$. *This expression can be found from the principle of conservation of mechanical energy.*

$$v = \pm\sqrt{\frac{k}{m}(A^2 - x^2)}$$

(13.6)

Speed as a function of position

The **period of a mass-spring system** depends on the mass (an inertial quantity) and the spring constant (an elastic quantity). The period of any object in simple harmonic motion is the time required to complete one full cycle of its motion.

$$T = 2\pi\sqrt{\frac{m}{k}}$$

(13.8)

The **harmonic frequency** (f) is the number of cycles or oscillations completed per unit time and is the reciprocal of the period. The units of frequency are hertz (Hz).

$$f = \frac{1}{T}$$

(13.9)

$$f = \frac{1}{2\pi}\sqrt{\frac{k}{m}}$$

(13.10)

Angular frequency (ω) is measured in radians per second.

$$\omega = 2\pi f = \sqrt{\frac{k}{m}}$$

(13.11)

The **initial conditions** illustrated in the figure are required for the particular form of Equations (13.14a), (3.14b), and (13.14c) as stated below. Note that the mass is at rest $(v = 0)$ and located at $x = A$ (the maximum displacement) when $t = 0$.

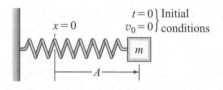

Position as a function of time

$$x = A\cos(2\pi f t)$$

(13.14a)

Velocity as a function of time

$$v = -A\omega\sin(2\pi f t)$$

(13.14b)

Acceleration as a function of time

$$a = -A\omega^2\cos(2\pi f t)$$

(13.14c)

The **period of a simple pendulum** depends only on its length and the local value of the acceleration due to gravity. To a good approximation, the period is independent of the amplitude (θ) within the range of small angular displacements.

$$T = 2\pi\sqrt{\frac{L}{g}} \qquad (13.15)$$

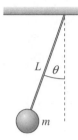

The **wave speed,** v, is the rate at which a disturbance or pulse moves along the direction of travel of the wave. *A traveling wave advances a distance equal to one wavelength in a time interval equal to one period.*

$$v = f\lambda \qquad (13.17)$$

The **wave speed in a stretched string** depends on the tension in the string and its linear density (mass per unit length). *For any mechanical wave, the speed depends only on the properties of the medium through which the wave travels.*

$$v = \sqrt{\frac{F}{\mu}} \qquad (13.18)$$

F = Tension in the string

REVIEW CHECKLIST

- Describe the general characteristics of a system in simple harmonic motion; define amplitude, period, frequency, and displacement.

- Define the following terms relating to wave motion: frequency, wavelength, velocity, and amplitude; express a given harmonic wave function in several alternative forms involving different combinations of the wave parameters: wavelength, period, phase velocity, angular frequency, and harmonic frequency.

- Given a specific wave function for a harmonic wave, obtain values for the characteristic wave parameters: A, ω, and f.

- Make calculations which involve the relationships between wave speed and the inertial and elastic characteristics of a string through which the disturbance is propagating.

- Define and describe the following wave-associated phenomena: superposition, phase, interference, and reflection.

SOLUTIONS TO SELECTED END-OF-CHAPTER PROBLEMS

1. A 0.60-kg block attached to a spring with force constant 130 N/m is free to move on a frictionless, horizontal surface as in Figure 13.7. The block is released from rest after the spring is stretched 0.13 m. At that instant, find (a) the force on the block and (b) its acceleration.

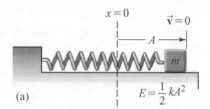

Figure 13.7 (a)

Solution

(a) Assuming that the spring obeys Hooke's law, the force it exerts on the block when the spring is stretched 0.13 m beyond its equilibrium length is

$$F_s = -kx = -(130 \text{ N/m})(+0.13 \text{ m}) = -17 \text{ N}$$

or

$$F_s = 17 \text{ N directed to the left (i.e., back toward the equilibrium position)} \qquad \Diamond$$

(b) The acceleration given the block by the spring force at the instant the block is released from rest will be

$$a = \frac{F_s}{m} = \frac{-17 \text{ N}}{0.60 \text{ kg}} = -28 \text{ m/s}^2$$

or

$$a = 28 \text{ m/s}^2 \text{ directed to the left (again, back toward the equilibrium position)} \qquad \Diamond$$

---

9. A slingshot consists of a light leather cup containing a stone. The cup is pulled back against two parallel rubber bands. It takes a force of 15 N to stretch either one of these bands 1.0 cm. (a) What is the potential energy stored in the two bands together when a 50-g stone is placed in the cup and pulled back 0.20 m from the equilibrium position? (b) With what speed does the stone leave the slingshot?

Solution

(a) Assume the rubber bands obey Hooke's law. Then, the force constant of either one of the bands is given by

$$k = \frac{F}{|\Delta x|} = \frac{15 \text{ N}}{1.0 \times 10^{-2} \text{ m}} = 1.5 \times 10^3 \text{ N/m}$$

Thus, the total elastic potential energy stored in the two bands when the cup is pulled back a distance $x_i = 0.20$ m is

$$PE_{s,i} = 2\left(PE_{\text{one band}} \right) = 2\left(\frac{1}{2}kx_i^2 \right) = \left(1.5 \times 10^3 \ \frac{\text{N}}{\text{m}} \right)(0.20 \text{ m})^2 = 60 \text{ J} \qquad \Diamond$$

(b) Consider the system consisting of the stone and the two bands. With the initial state being just before the cup is released and the final state being the instant the stone leaves the cup, assumed to occur at the equilibrium position (bands relaxed), conservation of mechanical energy, with $y_f = y_i$ (assuming the slingshot is held level) yields

$$KE_f + PE_{g,f} + PE_{s,f} = KE_i + PE_{g,i} + PE_{s,i}$$

which reduces to

$$\frac{1}{2} m_{stone} v_{max}^2 + \cancel{mgy_f} + 0 = 0 + \cancel{mgy_i} + PE_{s,i}$$

or

$$v_{max} = \sqrt{\frac{2(PE_{s,i})}{m_{stone}}} = \sqrt{\frac{2(60\text{ J})}{50 \times 10^{-3}\text{ kg}}} = 49\text{ m/s} \qquad \Diamond$$

13. A 10.0-g bullet is fired into, and embeds itself in, a 2.00-kg block attached to a spring with a force constant of 19.6 N/m and whose mass is negligible. How far is the spring compressed if the bullet has a speed of 300 m/s just before it strikes the block and the block slides on a frictionless surface? *Note:* You must use conservation of momentum in this problem. Why?

Solution

The impact between the bullet and the block is an inelastic collision, and it is very difficult to account for the various forms of energy present during the impact. This means that it is not practical to apply conservation of energy to any time interval that includes the moment of impact. We avoid this difficulty by using conservation of momentum from the instant just before the bullet strikes the block to the instant *immediately after impact* (i.e., before the block has had time to move and begin to compress the spring.) If v_0 is the speed of the bullet before impact and v_i is the speed of the block (with impeded bullet) immediately after impact, this gives

$$\left(M_{block} + m_{bullet} \right) v_i = m_{bullet} v_0$$

or

$$v_i = \frac{m_{bullet} v_0}{M_{block} + m_{bullet}} = \frac{\left(10.0 \times 10^{-3}\text{ kg}\right)\left(300\text{ m/s}\right)}{2.00\text{ kg} + 10.0 \times 10^{-3}\text{ kg}} = 1.49\text{ m/s}$$

Now, we apply conservation of energy from immediately after impact until the block comes to rest, obtaining

$$KE_f + PE_{g,f} + PE_{s,f} = KE_i + PE_{g,i} + PE_{s,i}$$

or

$$0 + \cancel{mgy_f} + \frac{1}{2}kx_f^2 = \frac{1}{2}\left(M_{block} + m_{bullet}\right)_i v_i^2 + \cancel{mgy_i} + 0$$

and giving the final distance the spring is compressed as

$$x_f = v_i \sqrt{\frac{M_{block} + m_{bullet}}{k}} = \left(1.49 \text{ m/s}\right)\sqrt{\frac{2.01 \text{ kg}}{19.6 \text{ N/m}}} = 0.477 \text{ m} \qquad \Diamond$$

---

19. At an outdoor market, a bunch of bananas is set into oscillatory motion with an amplitude of 20.0 cm on a spring with a force constant of 16.0 N/m. It is observed that the maximum speed of the bunch of bananas is 40.0 cm/s. What is the weight of the bananas in newtons?

Solution

The bananas have maximum speed as they pass through the equilibrium position, $x = 0$, where the elastic potential energy stored in the spring is zero. When the bananas are at maximum displacement from the equilibrium position, $x = \pm A$, they are momentarily at rest ($v = 0$). Since the Hooke's law type force exerted on the bananas by the spring is a conservative force, the total mechanical energy of the spring-bananas system is constant. Applying conservation of energy from when the bananas are at the equilibrium position to when they are at maximum displacement from equilibrium gives

$$KE_i + PE_{s,i} = KE_f + PE_{s,f} \implies \frac{1}{2}mv_{max}^2 + 0 = 0 + \frac{1}{2}kA^2$$

Thus, the mass of the bananas is

$$m = \frac{kA^2}{v_{max}^2} = \frac{\left(16.0 \text{ N/m}\right)\left(0.200 \text{ m}\right)^2}{\left(0.400 \text{ m/s}\right)^2} = 4.00 \text{ kg}$$

and their weight is

$$F_g = mg = \left(4.00 \text{ kg}\right)\left(9.80 \text{ m/s}^2\right) = 39.2 \text{ N} \qquad \Diamond$$

You should observe that in applying conservation of energy to this vertical spring-object system, we have ignored the gravitational potential energy of the object (bananas). It can be shown that it is valid to do this, and still have gravitational potential energy properly accounted for, if the zero point of the elastic potential energy is taken at the position where the object hangs in equilibrium on the end of the spring rather than at the position where the spring is completely relaxed and unstretched.

---

27. A cart of mass 250 g is placed on a frictionless horizontal air track. A spring having a spring constant of 9.5 N/m is attached between the cart and the left end of the track. When in equilibrium, the cart is located 12 cm from the left end of the track. If the cart is displaced 4.5 cm from its equilibrium position, find (a) the period at which it oscillates, (b) its maximum speed, and (c) its speed when it is 14 cm from the left end of the track.

Solution

(a) The period of an oscillating spring-object system is given by

$$T = \frac{2\pi}{\omega} = \frac{2\pi}{\sqrt{k/m}} = 2\pi\sqrt{\frac{m}{k}}$$

Thus, the period of this spring-cart system will be

$$T = 2\pi\sqrt{\frac{0.250 \text{ kg}}{9.5 \text{ N/m}}} = 1.0 \text{ s} \qquad \Diamond$$

(b) Since the cart is released from rest 4.5 cm from the equilibrium position, it will undergo simple harmonic motion with amplitude $A = 4.5$ cm about the equilibrium position. At the turning points $(x = \pm A)$, all of the energy is in the form of elastic potential energy, while at the equilibrium position $(x = 0)$, all of the energy is in the form of kinetic energy. Hence, $E = mv_{max}^2/2 = kA^2/2$, giving

$$v_{max} = A\sqrt{\frac{k}{m}} = (4.5 \text{ cm})\sqrt{\frac{9.5 \text{ N/m}}{0.250 \text{ kg}}} = 28 \text{ cm/s} = 0.28 \text{ m/s} \qquad \Diamond$$

(c) When the cart is located 14 cm from the left end of the track, it is at $x = 2.0$ cm from the equilibrium position. Conservation of energy, ignoring the constant gravitation potential energy of the cart on this level track, gives

$$E = KE + PE_s = 0 + PE_{s,\text{turning point}}, \qquad \text{or} \qquad \frac{1}{2}mv^2 + \frac{1}{2}kx^2 = 0 + \frac{1}{2}kA^2$$

and

$$v = \sqrt{\left(\frac{k}{m}\right)(A^2 - x^2)} = \sqrt{\left(\frac{9.5 \text{ N/m}}{0.250 \text{ kg}}\right)\left[(0.045 \text{ m})^2 - (0.020 \text{ m})^2\right]} = 0.25 \text{ m/s} \qquad \Diamond$$

33. Given that $x = A\cos(\omega t)$ is a sinusoidal function of time, show that v (velocity) and a (acceleration) are also sinusoidal functions of time. *Hint:* Use Equations 13.6 and 13.2.

Solution

The total energy of a simple harmonic oscillator is

$$E = KE + PE_s = \tfrac{1}{2}mv^2 + \tfrac{1}{2}kx^2$$

Since the force acting on the oscillating mass is a conservative force ($\vec{F}_s = -k\vec{x}$), the total energy is constant. At the maximum displacement from equilibrium ($x = A$), the speed is $v = 0$, so the total energy is $E = 0 + \tfrac{1}{2}kA^2 = \tfrac{1}{2}kA^2$. Thus, at arbitrary displacement x, we have $\tfrac{1}{2}mv^2 + \tfrac{1}{2}kx^2 = \tfrac{1}{2}kA^2$ which gives the velocity as

$$v = \pm\sqrt{\frac{k}{m}\left(A^2 - x^2\right)} = \pm\sqrt{\omega^2\left(A^2 - x^2\right)} = \pm\omega\sqrt{A^2 - x^2} \qquad \textbf{Equation (13.6)}$$

From Newton's second law, the acceleration is $\vec{a} = \vec{F}/m$. Thus, when $\vec{F} = \vec{F}_s = -k\vec{x}$, the acceleration is given by

$$a = -\frac{k}{m}x = -\omega^2 x. \qquad \textbf{Equation (13.2)}$$

For a simple harmonic oscillator, the position is a sinusoidal function of time, $x = A\cos(\omega t)$. The velocity is then

$$v = \pm\omega\sqrt{A^2 - x^2} = \pm\omega A\sqrt{1 - \cos^2(\omega t)} = \pm\omega A\sin(\omega t)$$

which is another sinusoidal function of time. $\quad\diamond$

The acceleration of the simple harmonic oscillator is given by

$$a = -\omega^2 x = -\left(\omega^2 A\right)\cos(\omega t), \text{ also a sinusoidal function of time} \quad\diamond$$

37. A pendulum clock that works perfectly on the Earth is taken to the Moon. (a) Does it run fast or slow there? (b) If the clock is started at 12:00 midnight, what will it read after one Earth day (24.0 h)? Assume that the free-fall acceleration on the Moon is 1.63 m/s^2.

Solution

(a) The period of a simple pendulum is $T = 2\pi\sqrt{\ell/g}$ where ℓ is the length of the pendulum and g is the local free-fall acceleration. Thus, if the pendulum clock is taken to the moon, where the free-fall acceleration is less than on Earth, the pendulum will have a longer period, causing the clock to run slow. ◊

(b) The ratio of the pendulum's period on the Moon to the period it had on Earth is

$$\frac{T_{Moon}}{T_{Earth}} = \frac{2\pi\sqrt{\ell/g_{Moon}}}{2\pi\sqrt{\ell/g_{Earth}}} = \sqrt{\frac{g_{Earth}}{g_{Moon}}} = \sqrt{\frac{9.80 \text{ m/s}^2}{1.63 \text{ m/s}^2}} = 2.45, \quad \text{or} \quad T_{Moon} = 2.45\, T_{Earth}$$

Since the frequency of the pendulum is the reciprocal of its period, the frequency of the clock's "ticks" on the Moon is

$$f_{Moon} = \frac{1}{T_{Moon}} = \frac{1}{2.45\left(1/f_{Earth}\right)}, \quad \text{or} \quad f_{Moon} = \frac{f_{Earth}}{2.45}$$

This means that when a time $\left(\Delta t\right)_{Earth}$ has elapsed on Earth clocks, the time that will appear to have passed according to the pendulum clock on the Moon is $\left(\Delta t\right)_{Moon} = \left(\Delta t\right)_{Earth}/2.45$. If the elapsed time on Earth is $\left(\Delta t\right)_{Earth} = 24.0$ h, the time that has passed according to the moon based clock is

$$\left(\Delta t\right)_{Moon} = \frac{24.0 \text{ h}}{2.45} = 9.80 \text{ h} = 9 \text{ h} + \left(0.80 \text{ h}\right)\left(\frac{60 \text{ min}}{1 \text{ h}}\right) = 9 \text{ h} + 48 \text{ min}$$

Since the pendulum clock on the Moon started with a reading of 12:00 midnight, its current reading will be 9:48 AM. ◊

48. Ocean waves are traveling to the east at 4.0 m/s with a distance of 20 m between crests. With what frequency do the waves hit the front of a boat (a) when the boat is at anchor and (b) when the boat is moving westward at 1.0 m/s?

Solution

(a) The relation between the wavelength, frequency, and wave speed is

$$v = \lambda f$$

The wavelength, λ, is the distance between successive crests of the wave, v is the speed at which the wave moves relative to the observer, and the frequency, f, is the number of crests that arrive at the observer each second.

The ocean waves have a distance $\lambda = 20$ m between crests. When the boat is at rest relative to the water, the speed of the waves relative to the boat is the same as their speed relative to the water, or $v = 4.0$ m/s. Thus, the frequency at which crests arrive at the boat is

$$f = \frac{v}{\lambda} = \frac{4.0 \text{ m/s}}{20 \text{ m}} = 0.20 \text{ s}^{-1} = 0.20 \text{ Hz} \qquad \Diamond$$

(b) If we take the positive direction for velocities to be eastward, the velocity of the waves relative to the water is $\vec{v}_{\text{wave,water}} = +4.0$ m/s and, from the discussion on relative velocity in Chapter 3 of the textbook, the velocity of the waves relative to the boat is

$$\vec{v}_{\text{wave,boat}} = \vec{v}_{\text{wave,water}} - \vec{v}_{\text{boat,water}}$$

If the boat is moving westward through the water at 1.0 m/s, then $\vec{v}_{\text{boat,water}} = -1.0$ m/s and the velocity of the wave relative to the boat is

$$\vec{v}_{\text{wave,boat}} = +4.0 \text{ m/s} - (-1.0 \text{ m/s}) = +5.0 \text{ m/s}$$

The frequency at which waves hit the front of the boat in this case is

$$f = \frac{v}{\lambda} = \frac{5.0 \text{ m/s}}{20 \text{ m}} = 0.25 \text{ s}^{-1} = 0.25 \text{ Hz} \qquad \Diamond$$

53. Transverse waves with a speed of 50.0 m/s are to be produced on a stretched string. A 5.00-m length of string with a total mass of 0.060 0 kg is used. (a) What is the required tension in the string? (b) Calculate the wave speed in the string if the tension is 8.00 N.

Solution

The mass per unit length of this string is

$$\mu = \frac{m}{\ell} = \frac{0.060\ 0\ \text{kg}}{5.00\ \text{m}} = 1.20 \times 10^{-2}\ \text{kg/m}$$

(a) The speed of transverse waves in a stretched string is given by

$$v = \sqrt{\frac{F}{\mu}}$$

where F is the tension in the string and μ is the mass per unit length of the string. Therefore, if the wave speed is to be 50.0 m/s in the given string, the required tension in the string is

$$F = \mu v^2 = \left(1.20 \times 10^{-2}\ \text{kg/m}\right)\left(50.0\ \text{m/s}\right)^2 = 30.0\ \text{N} \qquad \Diamond$$

(b) If the tension in the sting is now $F = 8.00$ N, the current speed of transverse waves in the string is

$$v = \sqrt{\frac{F}{\mu}} = \sqrt{\frac{8.00\ \text{N}}{1.20 \times 10^{-2}\ \text{kg/m}}} = 25.8\ \text{m/s} \qquad \Diamond$$

58. The elastic limit of a piece of steel wire is 2.70×10^9 Pa. What is the maximum speed at which transverse wave pulses can propagate along the wire without exceeding its elastic limit? (The density of steel is $7.86 \times 10^3 \ \text{kg}/\text{m}^3$.)

Solution

When a wire has tension F in it, the tensile stress is $Stress = F/A$, where A is the cross-sectional area of the wire. If the mass per unit length of the wire is $\mu = m/L$, the speed of transverse waves in the wire may be written as

$$v = \sqrt{\frac{F}{\mu}} = \sqrt{\frac{A \cdot Stress}{m/L}} = \sqrt{\frac{(A \cdot L)Stress}{m}}$$

The product $A \cdot L$, where A is the wire's cross-sectional area and L is its length, is just the volume V of material within the wire. Also, the mass of the wire is given by $m = \rho V$ where ρ is the density of the material making up the wire. The speed of transverse waves in the wire then becomes

$$v = \sqrt{\frac{(V)Stress}{\rho V}} = \sqrt{\frac{Stress}{\rho}}$$

If the maximum stress that can exist in a wire of density $\rho = 7.86 \times 10^3 \ \text{kg}/\text{m}^3$ is $(Stress)_{max} = 2.70 \times 10^9$ Pa, the maximum speed of transverse waves in this wire is

$$v_{max} = \sqrt{\frac{(Stress)_{max}}{\rho}} = \sqrt{\frac{2.70 \times 10^9 \ \text{Pa}}{7.86 \times 10^3 \ \text{kg}/\text{m}^3}} = 586 \ \text{m/s} \qquad \lozenge$$

71. A light balloon filled with helium of density 0.180 kg/m³ is tied to a light string of length $L = 3.00$ m. The string is tied to the ground, forming an "inverted" simple pendulum (Fig. P13.71a). If the balloon is displaced slightly from equilibrium, as in Figure P13.71b, show that the motion is simple harmonic and determine the period of the motion. Take the density of air to be 1.29 kg/m³. *Hint:* Use an analogy with the simple pendulum discussed in the text, and see Chapter 9.

Solution

At equilibrium, the string of the balloon is vertical. In the sketch at the right, the string (of length L) is at angle θ from the vertical, and the balloon is displaced distance $s = L\theta$ along a circular arc from the equilibrium position. In this orientation, the net force acting tangential to the circular path followed by the balloon is

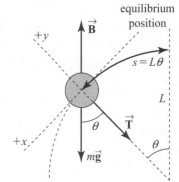

$$F_{net} = \Sigma F_x = -B\sin\theta + mg\sin\theta = -(B - mg)\sin\theta$$

The buoyant force acting on the balloon is

$$B = \begin{pmatrix} \text{weight of} \\ \text{displaced air} \end{pmatrix} = (\rho_{air}V)g$$

and the weight of the balloon is $mg = (\rho_{helium}V)g$, so the net force acting tangential to the path becomes

$$F_{net} = -\left[(\rho_{air}V)g - (\rho_{helium}V)g\right]\sin\theta = -(\rho_{air} - \rho_{helium})Vg\sin\theta$$

If the amplitude of the oscillations is small, then θ is always a small angle and we may use the approximation $\sin\theta \approx \theta = s/L$, (where θ is in radians). Then, the net force moving the balloon becomes

$$F_{net} = -(\rho_{air} - \rho_{helium})Vg\,s/L = -\left[\frac{(\rho_{air} - \rho_{helium})Vg}{L}\right]s = -k \cdot s$$

and we see that the net force is in the form of Hooke's law with an effective spring constant $k = (\rho_{air} - \rho_{helium})Vg/L$. Thus, the balloon will carry out simple harmonic motion with a period given by

$$T = 2\pi\sqrt{\frac{m}{k}} = 2\pi\sqrt{\frac{\rho_{helium}V}{(\rho_{air} - \rho_{helium})Vg/L}} = 2\pi\sqrt{\left(\frac{\rho_{helium}}{\rho_{air} - \rho_{helium}}\right)\frac{L}{g}} \qquad \Diamond$$

If $L = 3.00$ m, $\rho_{helium} = 0.180$ kg/m³, and $\rho_{air} = 1.29$ kg/m³, this period is

$$T = 2\pi\sqrt{\left(\frac{0.180 \text{ kg/m}^3}{1.29 \text{ kg/m}^3 - 0.180 \text{ kg/m}^3}\right)\frac{3.00 \text{ m}}{9.80 \text{ m/s}^2}} = 1.40 \text{ s} \qquad \Diamond$$

77. A large block P executes horizontal simple harmonic motion as it slides across a frictionless surface with a frequency $f = 1.50$ Hz. Block B rests on it, as shown in Figure P13.77, and the coefficient of static friction between the two is $\mu_s = 0.600$. What maximum amplitude of oscillation can the system have if block B is not to slip?

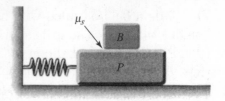

Figure P13.77

Solution

If block B is not to slip on block P, it must always have the same velocity (and hence, rate of change in velocity or acceleration) as does block P. From the free-body diagram of block B given at the right, we have

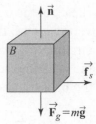

$$\Sigma F_y = ma_y = 0 \quad \text{or} \quad n = mg$$

and

$$\Sigma F_x = ma \quad \text{or} \quad a = \frac{f_s}{m}$$

Thus, the maximum acceleration block B can be given is determined by the maximum static friction force between it and block P. This is

$$\left(f_s\right)_{max} = \mu_s n = \mu_s mg$$

and the maximum acceleration of the blocks before slipping will occur is

$$a_{max} = \frac{\left(f_s\right)_{max}}{m} = \frac{\mu_s mg}{m} = \mu_s g$$

A system undergoing simple harmonic motion has its greatest acceleration at maximum displacement from equilibrium (that is, at $x = A$), and this acceleration is given by $a = A\omega^2 = A\left(2\pi f\right)^2$ where f is the frequency of vibration. Thus, the maximum amplitude of oscillation that does not cause slipping is

$$A_{max} = \frac{a_{max}}{\left(2\pi f\right)^2} = \frac{\mu_s g}{\left(2\pi f\right)^2} = \frac{(0.600)\left(9.80 \text{ m/s}^2\right)}{4\pi^2\left(1.50 \text{ s}^{-1}\right)^2} = 6.62 \times 10^{-2} \text{ m} = 6.62 \text{ cm} \qquad \lozenge$$

14

Sound

NOTES FROM SELECTED CHAPTER SECTIONS

14.1 Producing a Sound Wave

14.2 Characteristics of Sound Waves

Sound waves, which have as their source vibrating objects, are **longitudinal waves** traveling through a medium such as air. The particles of the medium oscillate back and forth along the direction in which the wave travels. This is in contrast to a transverse wave, in which the vibrations of the medium are at right angles to the direction of travel of the wave.

A sound wave traveling through air creates alternating regions of high and low molecular density and air pressure. A region of high density and air pressure is called a **compression** or **condensation;** a region of lower-than-normal density is referred to as a **rarefaction.** A sinusoidal curve can be used to represent a sound wave. *There are crests in the sinusoidal wave at the points where the sound wave has condensations, and troughs where the sound wave has rarefactions.*

14.4 Energy and Intensity of Sound Waves

The **intensity** of a wave is the rate at which sound energy flows through a unit area perpendicular to the direction of travel of the wave. The faintest sounds the human ear can detect have an intensity of about 1×10^{-12} W/m^2. This intensity is called the **threshold of hearing.** The loudest sounds the ear can tolerate, at the **threshold of pain,** have an intensity of about 1 W/m^2. The sensation of loudness is approximately logarithmic in the human ear, and the relative intensity of a sound is called the **intensity level** or **decibel level.**

14.5 Spherical and Plane Waves

The **intensity** of a spherical wave produced by a point source is proportional to the average power emitted and inversely proportional to the square of the distance from the source.

14.6 The Doppler Effect

In general, the Doppler effect is experienced whenever there is relative motion between source and observer. When the source and observer are moving toward each other, the frequency heard by the observer is higher than the frequency of the source. When the source and observer are moving away from each other, the observer hears a frequency lower than the source frequency.

14.8 Standing Waves

Standing waves can be set up in a string by the superposition of two wave trains traveling in opposite directions along the string. This can occur when waves reflected from one end of the string interfere with the incident wave train, under the condition that the wavelengths (and therefore frequencies) are properly matched to the length of the string. The string has a number of natural patterns of vibration, called **normal modes.** Each normal mode has a **characteristic frequency.** The lowest of these frequencies is called the **fundamental frequency,** which together with the higher frequencies form a **harmonic series.**

The figure on the right is a schematic representation of standing waves in a string of length L. In each case the envelope represents successive positions of the string during one complete cycle. Imagine that you observe the string vibrate while it is illuminated with a strobe light. The first three normal modes are shown. The points of zero displacement are called **nodes;** the points of maximum displacement are called **antinodes.**

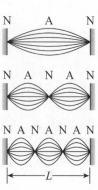

14.9 Forced Vibrations and Resonance

Consider a system (for example a mass-spring system) that has a natural frequency of vibration, f_O, and is driven or pushed back and forth with a periodic force whose frequency is f. This type of motion is referred to as a **forced vibration.** Its amplitude reaches a maximum when the frequency of the driving force equals the natural frequency of the system, f_O, called the **resonant frequency** of the system. Under this condition, the system is said to be in **resonance.**

14.10 Standing Waves in Air Columns

Sound sources can be used to produce **longitudinal standing waves** in air columns. The phase relationship between incident and reflected waves depends on whether or not the reflecting end of the air column is open or closed. This gives rise to two sets of possible standing wave conditions:

> In a **pipe open at both ends, all harmonics are present;** the natural frequencies of vibration form a series with frequencies equal to integral multiples of the fundamental.

> In a **pipe closed at one end and open at the other, only odd harmonics are present.**

14.11 Beats

Consider a type of interference effect that results from the superposition of two waves with slightly different frequencies. In this situation, at some fixed point the waves are periodically in and out of phase, corresponding to an alternation in time between constructive and destructive interference. A listener hears an alternation in loudness, known as beats. The number of beats per second, or beat frequency, equals the difference in frequency between the two sources.

EQUATIONS AND CONCEPTS

The **speed of sound in a fluid** depends on the value of the bulk modulus, B, (an elastic property) and the equilibrium density, ρ, (an inertial property) of the material through which it is traveling.

$$v = \sqrt{\frac{B}{\rho}} \qquad (14.1)$$

$$\text{where } B \equiv -\frac{\Delta P}{\Delta V / V} \qquad (14.2)$$

The **speed of sound in a solid** depends on the value of Young's modulus and the density of the material.

$$v = \sqrt{\frac{Y}{\rho}} \qquad (14.3)$$

The **speed of sound in air** depends on the Kelvin temperature as given by Equation (14.4). The speed of sound in air at 0°C (273 K) is 331 m/s.

$$v = (331 \text{ m/s}) \sqrt{\frac{T}{273 \text{ K}}} \qquad (14.4)$$

The **intensity of a wave** is the rate at which energy flows across a unit area in a plane perpendicular to the direction of travel of the wave. The SI units of intensity, I, are watts per square meter (W/m^2).

$$I \equiv \frac{\text{power}}{\text{area}} = \frac{\mathcal{P}}{A} \qquad (14.6)$$

The **decibel scale** is a logarithmic intensity scale. On this scale, the unit of sound intensity is the decibel, dB. The constant I_0 is a reference intensity chosen to coincide with the threshold of hearing. *Note: The term "log" indicates common logarithms (base 10) as opposed to "ln," which indicates natural logarithms (base e).* See an example of calculating the decibel level due to two sources sounded simultaneously in Suggestions, Skills, and Strategies.

$$\beta \equiv 10 \log \left(\frac{I}{I_0} \right) \qquad (14.7)$$

$$I_0 = 1.00 \times 10^{-12} \text{ W/m}^2$$

The **intensity of a spherical wave produced by a point source** is inversely proportional to the square of the distance from the source.

$$I = \frac{\mathcal{P}_{av}}{4 \pi r^2} \qquad (14.8)$$

The ratio of intensities at spherical surfaces of radii r_1 and r_2 is shown in Equation (14.9).

$$\frac{I_1}{I_2} = \frac{r_2^2}{r_1^2} \qquad (14.9)$$

Doppler effect (apparent change in frequency) is observed whenever there is relative motion between the source and the observer. Equation (14.12) is a general Doppler shift expression. **Proper use of signs for v_o and v_s is required depending on the relative motion of source and observer toward or away from each other!** See worked examples in Suggestions, Skills, and Strategies.

$$f_O = f_s \left(\frac{v + v_O}{v - v_s} \right) \qquad (14.12)$$

f_s = frequency of the source

f_O = observed frequency

v = velocity of sound (always positive)

v_O = velocity of observer

v_s = velocity of source

v_O and v_s are each measured relative to the medium in which the sound travels.

Apply the following sign rules when using Equation (14.12).

(1) If the source or observer is moving toward the other, enter its v with a positive sign.

(2) If the source or observer is moving away from the other, enter its v with a negative sign.

Normal modes of oscillation (a series of natural oscillations or "standing wave patterns") can be excited in a stretched string (fixed at both ends). Each mode corresponds to a characteristic frequency and wavelength.

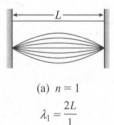 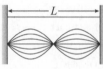 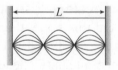

(a) $n = 1$ $\qquad$ (b) $n = 2$ $\qquad$ (c) $n = 3$

$$\lambda_1 = \frac{2L}{1} \qquad \lambda_2 = \frac{2L}{2} \qquad \lambda_3 = \frac{2L}{3}$$

Standing waves in a stretched string of length L fixed at both ends. The normal frequencies of vibration form a harmonic series: (a) the fundamental frequency, or first harmonic, (b) the second harmonic, and (c) the third harmonic.

The **observed frequencies** are integer multiples of the fundamental frequency (or first harmonic) f_1 corresponding to $n = 1$. This is the lowest frequency for which a standing wave is possible. The higher frequencies ($n = 2, 3, 4, \ldots$) form a harmonic series and can be expressed in terms of the wave speed and string length or in terms of the string tension and linear mass-density.

$$f_n = n \frac{v}{2L} \qquad n = 1, 2, 3, \ldots$$

$$f_n = \frac{n}{2L} \sqrt{\frac{F}{\mu}} \qquad n = 1, 2, 3, \ldots \qquad (14.17)$$

F = string tension

μ = linear mass-density

In an **"open"** pipe (open at both ends), the natural frequencies of vibration form a series in which all harmonics (all integer multiples of the fundamental) are present. *Note that this series of frequencies is the same as the series of frequencies produced in a string fixed at both ends.*

$$f_n = n\left(\frac{v}{2L}\right) \qquad n = 1, 2, 3, \ldots \quad (14.18)$$

1st harmonic

2nd harmonic

3rd harmonic

In a **"closed pipe"** (open at one end), only the odd harmonics (odd multiples of the fundamental) are possible.

$$f_n = n\left(\frac{v}{4L}\right) \qquad n = 1, 3, 5, \ldots \quad (14.19)$$

1st harmonic

3rd harmonic

5th harmonic

SUGGESTIONS, SKILLS, AND STRATEGIES

The Decibel Scale

When making calculations using Equation (14.7), which defines the intensity level of a sound wave on the decibel scale, the properties of logarithms must be kept clearly in mind. In order to determine the decibel level corresponding to two sources sounded simultaneously, you must first find the intensity, I, of each source in W/m^2, add these values, and then convert the resulting intensity to the decibel scale.

As an illustration of this technique, determine the dB level when two sounds with intensities of $\beta_1 = 40.0$ dB and $\beta_2 = 45.0$ dB are sounded together.

First solve Equation (14.7), $\beta = 10 \log (I/I_0)$, to find $I = I_0 10^{\beta/10}$.

Then for $\beta_1 = 40.0$ dB,

$$I_1 = (10^{-12}\,W/m^2)10^4 = 1.00 \times 10^{-8}\,W/m^2$$

For $\beta_2 = 45.0$ dB

$$I_2 = (10^{-12}\,W/m^2)10^{4.5} = 3.16 \times 10^{-8}\,W/m^2$$

and

$$I_{total} = I_1 + I_2 = 4.16 \times 10^{-8}\,W/m^2$$

Again using Equation (14.7), $\beta_{total} = 10 \log (I_{total}/I_0)$. So,

$$\beta_{total} = 10 \log \frac{\left(4.16 \times 10^{-8} \text{ W/m}^2\right)}{\left(1.00 \times 10^{-12} \text{ W/m}^2\right)} = 10 \log (4.16 \times 10^4) = 46.2 \text{ dB}$$

The intensity level of the combined sources is 46.2 dB (not 40 dB + 45 dB = 85 dB).

Doppler Effect

The most likely error in using Equation (14.12) to calculate the Doppler frequency shift due to relative motion between a sound source and an observer is due to using the incorrect algebraic sign for the velocity of either the observer or the source. Remember the following relationship between relative velocity of source and observer and the corresponding Doppler frequency shift: The word **toward** is associated with an **increase** in frequency and the words **away from** are associated with a **decrease** in frequency.

Consider the following examples, given a source frequency of 300 Hz and the speed of sound in air of 343 m/s.

(1) Source moving with a speed of 40 m/s toward a fixed observer. In this case, $v_O = 0$ and $v_S = +40$ m/s. Substituting into Equation (14.12),

$$f_O = \left(\frac{v + v_O}{v - v_S}\right) f_S = \left(\frac{343 \text{ m/s} + 0}{343 \text{ m/s} - 40 \text{ m/s}}\right)(300 \text{ Hz}) = 340 \text{ Hz}$$

(2) Source moving away from a fixed observer with a speed of 40 m/s. In this case, $v_O = 0$ and $v_S = -40$ m/s. Substituting into Equation (14.12),

$$f_O = \left(\frac{v + v_O}{v - v_S}\right) f_S = \left(\frac{343 \text{ m/s} + 0}{343 \text{ m/s} - (-40 \text{ m/s})}\right)(300 \text{ Hz}) = 269 \text{ Hz}$$

(3) Observer moving with speed of 40 m/s toward a fixed source: $v_O = +40$ m/s and $v_S = 0$. Substituting into Equation (14.12),

$$f_O = \left(\frac{v + v_O}{v - v_S}\right) f_S = \left(\frac{343 \text{ m/s} + 40 \text{ m/s}}{343 \text{ m/s} - 0}\right)(300 \text{ Hz}) = 335 \text{ Hz}$$

(4) Observer moving with speed 40 m/s away from a fixed source: $v_O = -40$ m/s and $v_S = 0$. Substituting into Equation (14.12),

$$f_O = \left(\frac{v + v_O}{v - v_S}\right) f_S = \left(\frac{343 \text{ m/s} + (-40 \text{ m/s})}{343 \text{ m/s} - 0}\right)(300 \text{ Hz}) = 265 \text{ Hz}$$

As a check on your calculated value of f_O, remember:

- When the relative motion is source or observer toward the other, $f_O > f_S$.

- When the relative motion is source or observer away from the other, $f_O < f_S$.

REVIEW CHECKLIST

- Describe the harmonic displacement and pressure variation as functions of time and position for a harmonic sound wave.

- Calculate the speed of sound in various media in terms of appropriate elastic properties (these can include bulk modulus, Young's modulus, and the pressure-volume relationships of an ideal gas) and the corresponding inertial properties (usually the mass density).

- Understand the basis of the logarithmic intensity scale (decibel scale) and convert intensity values (given in W/m^2) to loudness levels on the dB scale. Determine the intensity ratio for two sound sources whose decibel levels are known. Calculate the intensity of a point source wave at a given distance from the source. Remember that you must use common logarithms (base 10) in Equation (14.7).

- Describe the various situations under which a Doppler shifted frequency is produced. Calculate the apparent frequency for a given actual frequency for each of the various possible relative motions between source and observer.

- Describe in both qualitative and quantitative terms the conditions that produce standing waves in a stretched string and in an open or closed air column pipe.

SOLUTIONS TO SELECTED END-OF-CHAPTER PROBLEMS

7. You are watching a pier being constructed on the far shore of a saltwater inlet when some blasting occurs. You hear the sound in the water 4.50 s before it reaches you through the air. How wide is the inlet? *Hint*: See Table 14.1. Assume the air temperature is 20°C.

Solution

From Table 14.1, the speed of sound in the saltwater is $v_w = 1530$ m/s. At $T = 20°C = 293$ K, the speed of sound in air is

$$v_a = (331 \text{ m/s})\sqrt{\frac{T}{273 \text{ K}}} = (331 \text{ m/s})\sqrt{\frac{293 \text{ K}}{273 \text{ K}}} = 343 \text{ m/s}$$

If d is the width of the inlet, the time required for sound traveling in air to cross the inlet is $t_a = d/v_a$. Likewise, the required transit time for sound traveling through seawater is $t_w = d/v_w$.

Since it is given that $t_a = t_w + 4.50$ s, we have

$$\frac{d}{v_a} = \frac{d}{v_w} + 4.50 \text{ s}, \quad \text{or} \quad d = (4.50 \text{ s})\left(\frac{v_w v_a}{v_w - v_a}\right)$$

Therefore, the width of the inlet must be

$$d = (4.50 \text{ s})\left[\frac{(1530 \text{ m/s})(343 \text{ m/s})}{(1530 - 343) \text{ m/s}}\right] = 1.99 \times 10^3 \text{ m} = 1.99 \text{ km} \qquad \lozenge$$

13. A person wears a hearing aid that uniformly increases the intensity level of all audible frequencies of sound by 30.0 dB. The hearing aid picks up sound having a frequency of 250 Hz at an intensity of 3.0×10^{-11} W/m^2. What is the intensity delivered to the eardrum?

Solution

The decibel level of a sound of intensity I is given by $\beta = 10\log(I/I_0)$, where $I_0 = 1.0 \times 10^{-12}$ W/m^2 is the intensity of sound at the threshold of hearing. If the incident sound has intensity $I_1 = 3.0 \times 10^{-11}$ W/m^2, the sound level reaching the ear drum without the hearing aid would be

$$\beta_1 = 10\log\left(\frac{I_1}{I_0}\right) = 10\log\left(\frac{3.0 \times 10^{-11} \text{ W/m}^2}{1.0 \times 10^{-12} \text{ W/m}^2}\right) = 14.8 \text{ dB}$$

Since the hearing aid increases the sound level of all audible frequencies by 30.0 dB, the sound level reaching the eardrum with the hearing aid in place is

$$\beta_2 = \beta_1 + 30.0 \text{ dB} = 14.8 \text{ dB} + 30.0 \text{ dB} = 44.8 \text{ dB}$$

and the intensity of this sound is

$$I_2 = I_0 \cdot 10^{\beta_2/10} = \left(1.0 \times 10^{-12} \text{ W/m}^2\right) \cdot 10^{4.48} = 3.0 \times 10^{-8} \text{ W/m}^2 \qquad \Diamond$$

As an alternate approach to this problem, observe using the property of logarithms that $\log A - \log B = \log(A/B)$ allows one to write the difference in the decibel level of two sounds as

$$\beta_2 - \beta_1 = 10\log\left(\frac{I_2}{I_0}\right) - 10\log\left(\frac{I_1}{I_0}\right) = 10\log\left(\frac{I_2}{I_0} \cdot \frac{I_0}{I_1}\right) = 10\log\left(\frac{I_2}{I_1}\right)$$

Since the hearing aid increases the decibel level by 30 dB, we have $\beta_2 - \beta_1 = 30$ dB, and the ratio of the intensities reaching the eardrum with and without the hearing aid in place is

$$\frac{I_2}{I_1} = 10^{(\beta_2 - \beta_1)/10} = 10^{(30 \text{ dB})/10} = 10^3$$

giving

$$I_2 = 10^3 \cdot I_1 = 10^3 \left(3.0 \times 10^{-11} \text{ W/m}^2\right) = 3.0 \times 10^{-8} \text{ W/m}^2 \qquad \Diamond$$

21. Show that the difference in decibel levels β_1 and β_2 of a sound source is related to the ratio of its distances r_1 and r_2 from the receivers by the formula

$$\beta_2 - \beta_1 = 20\log\left(\frac{r_1}{r_2}\right).$$

Solution

Assuming the sound source radiates equally well in all directions, the wave fronts are spheres centered on the source. The intensity of the sound from such a source varies inversely as the square of the distance from the source, or

$$I = \frac{\mathcal{P}_{av}}{4\pi r^2}$$

where $\mathcal{P}_{av}$ is the average power emitted by the source and r is the distance of the observer from the source.

Since the decibel level of a sound of intensity I is defined as $\beta = 10\log(I/I_0)$, where I_0 is a constant, the difference in the decibel levels at distances r_1 and r_2 from a source is given by

$$\beta_2 - \beta_1 = 10\log\left(\frac{I_2}{I_0}\right) - 10\log\left(\frac{I_1}{I_0}\right) = 10\log\left(\frac{I_2}{I_0}\cdot\frac{I_0}{I_1}\right) = 10\log\left(\frac{I_2}{I_1}\right)$$

where we have used the fact that $\log A - \log B = \log(A/B)$. In terms of the distances from the source, this becomes

$$\beta_2 - \beta_1 = 10\log\left(\frac{\mathcal{P}_{av}}{4\pi r_2^2}\cdot\frac{4\pi r_1^2}{\mathcal{P}_{av}}\right) = 10\log\left(\frac{r_1^2}{r_2^2}\right) = 10\log\left[\left(\frac{r_1}{r_2}\right)^2\right]$$

Finally, making use of the property of logarithms that $\log A^m = m\log A$, we obtain

$$\beta_2 - \beta_1 = 10\log\left[\left(\frac{r_1}{r_2}\right)^2\right] = 10\left[2\log\left(\frac{r_1}{r_2}\right)\right], \quad \text{or} \quad \beta_2 - \beta_1 = 20\log\left(\frac{r_1}{r_2}\right) \qquad \Diamond$$

25. Two trains on separate tracks move toward each other. Train 1 has a speed of 130 km/h, train 2 a speed of 90.0 km/h. Train 2 blows its horn, emitting a frequency of 500 Hz. What is the frequency heard by the engineer on train 1?

Solution

When a source moving with velocity v_S emits sound having frequency f_S, the frequency f_O detected by an observer moving with velocity v_O is

$$f_O = f_S \left(\frac{v + v_O}{v - v_S} \right)$$

Here, v is the velocity of sound in the propagating medium. Also, v_S is positive if the source moves toward the observer and negative if the source moves away from the observer. Similarly, v_O is positive when the observer moves toward the source and negative if the observer moves away from the source.

In this case, train 2 is the source of a sound having frequency $f_S = 500$ Hz and it is moving toward the observer (train 1) at 90 km/h. Thus, $v_S = +90$ km/h. The observer is on train 1 and moves toward the source (train 2) at 130 km/h, so $v_O = +130$ km/h. The sound travels through air with a speed of

$$v = \left(345 \ \frac{\text{m}}{\text{s}} \right) \left(\frac{1 \ \text{km}}{10^3 \ \text{m}} \right) \left(\frac{3\,600 \ \text{s}}{1 \ \text{h}} \right) = 1\,240 \ \text{km/h}$$

The frequency detected by the observer in train 1 is then

$$f_O = (500 \ \text{Hz}) \left(\frac{1\,240 \ \text{km/h} + 130 \ \text{km/h}}{1\,240 \ \text{km/h} - 90.0 \ \text{km/h}} \right) = 596 \ \text{Hz} \qquad \Diamond$$

29. A tuning fork vibrating at 512 Hz falls from rest and accelerates at 9.80 m/s². How far below the point of release is the tuning fork when waves of frequency 485 Hz reach the release point? Take the speed of sound in air to be 340 m/s.

Solution

With $v_S = -|v_S|$ (negative since the source (tuning fork) moves *away* from the observer), and $v_O = 0$ (since the observer is stationary), the Doppler-shifted frequency $f_O = 485$ Hz detected by the observer from the source of frequency $f_S = 512$ Hz is given by the relation $f_O = f_S (v + v_O)/(v - v_S)$ as

$$\frac{f_O}{f_S} = \frac{485 \text{ Hz}}{512 \text{ Hz}} = \frac{340 \text{ m/s} + 0}{340 \text{ m/s} - \left(-|v_S|\right)}$$

yielding

$$|v_S| = 340 \text{ m/s}\left[\left(\frac{512 \text{ Hz}}{485 \text{ Hz}}\right) - 1\right] = 18.9 \text{ m/s}$$

Thus, the first step in determining where the tuning fork is located when the observer hears the 485 Hz sound is to find how far the tuning fork must fall (starting from rest, with downward acceleration $g = 9.80$ m/s²) before it reaches the required speed of $|v_S| = 18.9$ m/s. Taking downward as positive, and using $v_y^2 = v_{0y}^2 + 2a_y(\Delta y)$, this displacement is found to be

$$(\Delta y)_1 = \frac{|v_S|^2}{2g} = \frac{(18.9 \text{ m/s})^2}{2(9.80 \text{ m/s}^2)} = 18.3 \text{ m}$$

The time required for the sound emitted by the tuning fork at this location to travel back to the release point is

$$\Delta t = \frac{(\Delta y)_1}{v_{\text{sound}}} = \frac{18.3 \text{ m}}{340 \text{ m/s}} = 0.0538 \text{ s}$$

During this time interval, the tuning fork will drop an additional distance of

$$(\Delta y)_2 = |v_S|(\Delta t) + \frac{1}{2} g(\Delta t)^2$$

$$= (18.9 \text{ m/s})(0.0538 \text{ s}) + \frac{1}{2}(9.80 \text{ m/s}^2)(0.0538 \text{ s})^2 = 1.03 \text{ m}$$

Thus, at the instant the observer at the release point hears the 485 Hz sound, the tuning fork is located

$$(\Delta y)_{\text{total}} = (\Delta y)_1 + (\Delta y)_2 = 18.3 \text{ m} + 1.03 \text{ m} = 19.3 \text{ m below that point} \qquad \lozenge$$

35. The ship in Figure P14.35 travels along a straight line parallel to the shore and 600 m from it. The ship's radio receives simultaneous signals of the same frequency from antennas A and B. The signals interfere constructively at point C, which is equidistant from A and B. The signal goes through the first minimum at point D. Determine the wavelength of the radio waves.

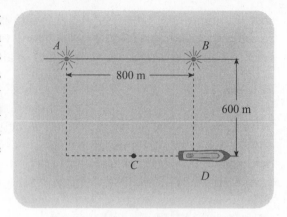

Figure P14.35

Solution

When the ship is at point D, the distance to the ship from the antenna at point A is given by the Pythagorean theorem as

$$d_A = \sqrt{(800 \text{ m})^2 + (600 \text{ m})^2} = 1\,000 \text{ m}$$

The distance to the ship from the antenna at point B is 600 m. Thus, the difference in the path lengths the radio waves must travel from the two antennas to the ship is

$$\Delta d = d_A - d_B = 1\,000 \text{ m} - 600 \text{ m} = 400 \text{ m}$$

If the two signals reaching the ship are to interfere destructively and yield a minimum of intensity, the difference in path lengths must be an odd number of half-wavelengths, or $\Delta d = \left(n + \frac{1}{2}\right)\lambda$ where $(n = 0, 1, 2, 3, \ldots)$. Since the ship is located at the position of the first minimum in the interference pattern produced by the waves from the two sources, n must have its lowest possible value, or $n = 0$. Hence, $\Delta d = \lambda/2$ and the wavelength of the radio waves must be

$$\lambda = 2\Delta d = 2(400 \text{ m}) = 800 \text{ m} \qquad \Diamond$$

41. Two speakers are driven by a common oscillator at 800 Hz and face each other at a distance of 1.25 m. Locate the points along a line joining the speakers where relative minima of the amplitude of the pressure would be expected. (Use $v = 343$ m/s.)

Solution

Since the speakers are driven by a common oscillator, they must vibrate in phase with each other. Thus, the point halfway between them (being equidistant from the two sources) must be an antinode in any standing wave pattern formed.

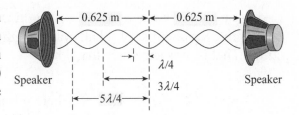

Relative minima (i.e., nodes) are located on either side of this central antinode at a distance of

$$\frac{\lambda}{4} = \frac{1}{4}\left(\frac{v}{f}\right) = \frac{343 \text{ m/s}}{4(800 \text{ Hz})} = 0.107 \text{ m from the midpoint}$$

This means that they are located at distances of

$$d = 0.625 \text{ m} \pm 0.107 \text{ m} = 0.518 \text{ m and } 0.732 \text{ m from either speaker} \qquad \lozenge$$

A second pair of nodes will be found at a distance of $3\lambda/4 = 3(0.107$ m$)$ on either side of the midpoint. The positions of these nodes are then located at distances of

$$d = 0.625 \text{ m} \pm 3(0.107 \text{ m}) = 0.303 \text{ m and } 0.947 \text{ m from either speaker} \qquad \lozenge$$

Finally a pair of nodes will be found at a distance of $5\lambda/4 = 5(0.107$ m$)$ on either side of the midpoint. These points are located at distances of

$$d = 0.625 \text{ m} \pm 5(0.107 \text{ m}) = 0.089 \text{ m and } 1.16 \text{ m from either speaker} \qquad \lozenge$$

47. A 60.00-cm guitar string under a tension of 50.000 N has a mass per unit length of 0.100 00 g/cm. What is the highest resonant frequency that can be heard by a person capable of hearing frequencies up to 20 000 Hz?

Solution

In order to produce resonance in a string fixed at both ends, the length of the string must be any integer multiple of a half-wavelength of the transverse waves traveling along the string. That is, it is necessary that

$$L = n\left(\frac{\lambda_n}{2}\right), \quad \text{or} \quad \lambda_n = \frac{2L}{n} \text{ where } n \text{ is any positive integer}$$

Thus, the resonant frequencies of the string are

$$f_n = \frac{v}{\lambda_n} = n\left(\frac{v}{2L}\right)$$

where v is the speed of transverse waves in the string. If a string having mass per unit length

$$\mu = \left(0.100\ 00\ \frac{g}{cm}\right)\left(\frac{10^2\ cm}{1\ m}\right)\left(\frac{1\ kg}{10^3\ g}\right) = 1.000\ 0 \times 10^{-2}\ kg/m$$

is under a tension of $F = 50.000$ N, the speed of transverse waves in it is

$$v = \sqrt{\frac{F}{\mu}} = \sqrt{\frac{50.000\ N}{1.000\ 0 \times 10^{-2}\ kg/m}} = 70.711\ m/s$$

If the length of the string is $L = 60.00$ cm, the highest resonance with frequency less than or equal to 20 000 Hz (and therefore audible to the listener) is

$$n \le f\left(\frac{2L}{v}\right) = \left(20\ 000\ s^{-1}\right)\left[\frac{2(0.600\ 0\ m)}{70.711\ m/s}\right] = 339.4$$

Since n must have an integer value, we have $n = 339$ and the frequency at which this resonance occurs is

$$f_{339} = n\left(\frac{v}{2L}\right) = 339\left(\frac{70.711\ m/s}{2(0.600\ 0\ m)}\right) = 19\ 980\ Hz = 19.98\ kHz \qquad \lozenge$$

59. A student holds a tuning fork oscillating at 256 Hz. He walks toward a wall at a constant speed of 1.33 m/s. (a) What beat frequency does he observe between the tuning fork and its echo? (b) How fast must he walk away from the wall to observe a beat frequency of 5.00 Hz?

Solution

Initially, the student serves as a moving source $\left(v'_S = v_{student}\right)$ of sound having $f_S = 256$ Hz. The wall acts as a stationary $\left(v'_O = 0\right)$ receiver or observer of this sound. The sound wave received and reflected by the wall has frequency

$$f_{echo} = f'_O = f_S\left(\frac{v + v'_O}{v - v'_S}\right) = f_S\left(\frac{v}{v - v_{student}}\right)$$

where $v = 345$ m/s is the speed of sound in air.

At the reflection, the wall serves as a stationary source $\left(v_S = 0\right)$ of sound having frequency f_{echo}, and the student is a moving observer $\left(v_O = v_{student}\right)$ of this sound. The frequency of the echo heard by the student is

$$f_O = f_{echo}\left(\frac{v + v_O}{v - v_S}\right) = \left[f_S\left(\frac{v}{v - v_{student}}\right)\right]\left(\frac{v + v_{student}}{v}\right) = f_S\left(\frac{v + v_{student}}{v - v_{student}}\right)$$

The beat frequency heard by the student is then

$$f_{beat} = \left|f_O - f_S\right| = \left|f_S\left(\frac{v + v_{student}}{v - v_{student}}\right) - f_S\right| = 2f_S\left|\frac{v_{student}}{v - v_{student}}\right|$$

(a) If the student *approaches* the wall at speed $\left|v_{student}\right| = 1.33$ m/s, the needed velocity is $v_{student} = +1.33$ m/s, and the beat frequency is

$$f_{beat} = 2f_S\left|\frac{v_{student}}{v - v_{student}}\right| = 2(256\text{ Hz})\left|\frac{+1.33\text{ m/s}}{345\text{ m/s} - 1.33\text{ m/s}}\right| = 1.98\text{ Hz} \qquad \lozenge$$

(b) If the student *moves away* from the wall at speed $\left|v_{student}\right|$, the needed velocity is $v_{student} = -\left|v_{student}\right|$. If the student is to hear $f_{beat} = 5.00$ Hz, then

$$2f_S\left|\frac{-\left|v_{student}\right|}{v + \left|v_{student}\right|}\right| = 5.00\text{ Hz} \qquad \text{or} \qquad \frac{2f_S\left|v_{student}\right|}{v + \left|v_{student}\right|} = 5.00\text{ Hz}$$

This yields

$$\left|v_{student}\right| = v\left(\frac{5.00\text{ Hz}}{2f_S - 5.00\text{ Hz}}\right) = \left(345\frac{\text{m}}{\text{s}}\right)\left[\frac{5.00\text{ Hz}}{2(256\text{ Hz}) - 5.00\text{ Hz}}\right] = 3.40\text{ m/s} \qquad \lozenge$$

66. A student uses an audio oscillator of adjustable frequency to measure the depth of a water well. He reports hearing two successive resonances at 52.0 Hz and 60.0 Hz. If the speed of sound is 345 m/s, how deep is the well?

Solution

The well will act as a pipe closed at the lower end and open at the upper end. At resonance, the sound waves form a standing wave pattern with a node at the bottom of the well and an antinode at the top of the well. Since the distance from a node to the nearest antinode is a quarter wavelength, The longest resonant wavelength (i.e., the fundamental) which can meet the requirements of a node at the bottom and an antinode at the top is $\lambda_1 = 4L$, where L is the depth of the well. The lowest resonant frequency of the well is then $f_1 = v/\lambda_1 = v/4L$, where v is the speed of sound in air.

In a standing wave pattern, the distance between adjacent nodes is a half-wavelength. Thus, when the well is resonating in a higher harmonic, the distance from the node at the bottom of the well and the uppermost node must be an integer multiple of $\lambda/2$. The distance from the uppermost node on to the antinode at the top of the well is $\lambda/4$, so the total length of the well is

$$L = n\frac{\lambda}{2} + \frac{\lambda}{4} = (2n+1)\frac{\lambda}{4} \quad (n = 0, 1, 2, 3, \ldots)$$

or equivalently,

$$L = (2n-1)\frac{\lambda}{4} \quad (n = 1, 2, 3, \ldots)$$

The resonant wavelengths of this pipe closed at one end and open at the other are

$$\lambda_n = \frac{4L}{(2n-1)} \quad n = 1, 2, 3, \ldots$$

and the resonant frequencies are

$$f_n = \frac{v}{\lambda_n} = (2n-1)\frac{v}{4L} = (2n-1)f_1, \quad n = 1, 2, 3, \ldots$$

Observe that this says the possible resonant frequencies of the well are f_1, $3f_1$, $5f_1$, $7f_1$, $9f_1$, ... That is, only the *odd integer* multiples of the fundamental frequency are resonant frequencies for a pipe closed at one end and open at the other, and the difference between any two successive resonant frequencies is $2f_1$. Thus, if the well is observed to have successive resonant frequencies of 52.0 Hz and 60.0 Hz, we know that

$$2f_1 = 2\left(\frac{v}{4L}\right) = 60.0 \text{ Hz} - 52.0 \text{ Hz} = 8.00 \text{ Hz}$$

and the depth of the well is

$$L = 2\left[\frac{v}{4(8.00 \text{ Hz})}\right] = \frac{345 \text{ m/s}}{16.0 \text{ Hz}} = 21.6 \text{ m} \qquad \Diamond$$

71. On a workday the average decibel level of a busy street is 70 dB, with 100 cars passing a given point every minute. If the number of cars is reduced to 25 every minute on a weekend, what is the decibel level of the street?

Solution

The decibel level of a sound is given by $\beta = 10\log(I/I_0)$, where I is the intensity of the sound and $I_0 = 1.0 \times 10^{-12}$ W/m^2 is a reference intensity.

On the weekend, there are one-fourth as many cars passing per minute as on a week day. Thus, the expected sound intensity, I_2, on the weekend should be one-fourth the sound intensity, I_1, on a week day. The difference in the decibel levels on a week day and on the weekend will be

$$\beta_1 - \beta_2 = 10\log\left(\frac{I_1}{I_0}\right) - 10\log\left(\frac{I_2}{I_0}\right) = 10\log\left(\frac{I_1}{I_2}\right) = 10\log(4) = 6 \text{ dB}$$

The decibel level on the weekend is

$$\beta_2 = \beta_1 - 6 \text{ dB} = 70 \text{ dB} - 6 \text{ dB} = 64 \text{ dB} \qquad \Diamond$$

As an alternate approach, we compute the sound intensity on a week day as

$$I_1 = I_0 \cdot 10^{\beta_1/10} = I_0 \cdot 10^{70/10} = \left(1.0 \times 10^{-12} \text{ W/m}^2\right) \cdot 10^7 = 1.0 \times 10^{-5} \text{ W/m}^2$$

The expected sound intensity on a weekend, when one-fourth as many cars are passing each minute, is

$$I_2 = \frac{I_1}{4} = \frac{1.0 \times 10^{-5} \text{ W/m}^2}{4} = 2.5 \times 10^{-6} \text{ W/m}^2$$

and the expected decibel level on the weekend is therefore

$$\beta_2 = 10\log\left(\frac{I_2}{I_0}\right) = 10\log\left(\frac{2.5 \times 10^{-6} \text{ W/m}^2}{1.0 \times 10^{-12} \text{ W/m}^2}\right) = 10(6.40) = 64 \text{ dB} \qquad \Diamond$$

74. A student stands several meters in front of a smooth reflecting wall, holding a board on which a wire is fixed at each end. The wire, vibrating in its third harmonic, is 75.0 cm long, has a mass of 2.25 g, and is under a tension of 400 N. A second student, moving towards the wall, hears 8.30 beats per second. What is the speed of the student approaching the wall? Use 340 m/s as the speed of sound in air.

Solution

When the wire, fixed at both ends and of length $L = 0.750$ m, vibrates in its third harmonic, the wavelength of the standing wave in the wire is

$$\lambda = \frac{2L}{3} = \frac{2(0.750 \text{ m})}{3} = 0.500 \text{ m}$$

The speed of transverse waves in the wire is given by

$$v_{\text{wire}} = \sqrt{\frac{F}{\mu}} = \sqrt{\frac{F}{m/L}} = \sqrt{\frac{FL}{m}} = \sqrt{\frac{(400 \text{ N})(0.750 \text{ m})}{2.25 \times 10^{-3} \text{ kg}}} = 365 \text{ m/s}$$

The frequency at which the wire vibrates, and hence the frequency of the sound it generates in the surrounding air, is

$$f_S = \frac{v_{\text{wire}}}{\lambda} = \frac{365 \text{ m/s}}{0.500 \text{ m}} = 730 \text{ Hz}$$

Since both the wire and the wall are stationary, the frequency of the reflected wave from the wall is the same as the frequency of the wave coming directly from the wire. The student will then hear sound from two stationary sources ($v_S = 0$), both emitting a frequency $f_S = 730$ Hz. Walking at speed $|v_{\text{student}}|$, the student walks *away* from the wire (source 1) with velocity $v_{O,1} = -|v_{\text{student}}|$, and *toward* the reflecting wall (source 2) with velocity $v_{O,2} = +|v_{\text{student}}|$. Thus, the sound coming directly from the wire to the student is Doppler shifted to a lower frequency while the sound reflecting from the wall is Doppler shifted to a higher frequency. The beat frequency the student will detect while hearing these two sounds simultaneously is

$$f_{\text{beat}} = f_{O,2} - f_{O,1} = f_S \left(\frac{v_{\text{air}} + v_{O,2}}{v_{\text{air}} - v_{S,2}} \right) - f_S \left(\frac{v_{\text{air}} + v_{O,1}}{v_{\text{air}} - v_{S,1}} \right) = f_S \left(\frac{v_{\text{air}} + (v_{\text{student}})}{v_{\text{air}} - 0} \right) - f_S \left(\frac{v_{\text{air}} - (v_{\text{student}})}{v_{\text{air}} - 0} \right)$$

or

$$f_{\text{beat}} = \frac{2 f_S |v_{\text{student}}|}{v_{\text{air}}}$$

If it is observed that $f_{\text{beat}} = 8.30$ Hz, the speed of the walking student is

$$|v_{\text{student}}| = \frac{v_{\text{air}} f_{\text{beat}}}{2 f_S} = \frac{(340 \text{ m/s})(8.30 \text{ Hz})}{2(730 \text{ Hz})} = 1.93 \text{ m/s} \qquad \Diamond$$